# SpringerBriefs in Microbiology

Extremophilic Bacteria

**Series editors**

Sonia M. Tiquia-Arashiro, Dearborn, MI, USA
Melanie Mormile, Rolla, MO, USA

More information about this series at http://www.springer.com/series/11917

Sonia M. Tiquia-Arashiro

# Thermophilic Carboxydotrophs and their Applications in Biotechnology

 Springer

Sonia M. Tiquia-Arashiro
Department of Natural Sciences
University of Michigan
Dearborn, MI
USA

ISSN 2191-5385           ISSN 2191-5393   (electronic)
ISBN 978-3-319-11872-7   ISBN 978-3-319-11873-4   (eBook)
DOI 10.1007/978-3-319-11873-4

Library of Congress Control Number: 2014951696

Springer Cham Heidelberg New York Dordrecht London

Springer is part of Springer Science+Business Media (www.springer.com)

*This book is dedicated to my husband,*
*Peter whose unlimited devotion*
*and support made this book project possible*

*To the reader...*

*I hope that you have at least half*
*as much fun in reading this book as*
*I've had in writing it.*

# Contents

1  Introduction . . . . . . . . . . . . . . . . . . . . . . . . . . . . . . . . . . . . . .    1

2  Microbial CO Metabolism  . . . . . . . . . . . . . . . . . . . . . . . . . . . . .    5

3  CO-oxidizing Microorganisms . . . . . . . . . . . . . . . . . . . . . . . . . .   11
   3.1  CO-utilizing Hydrogenogenic Bacteria and Archaea . . . . . . . . . .   12
        3.1.1  CO-utilizing Facultatively Anaerobic Bacteria . . . . . . . . .   15
        3.1.2  CO-utilizing Obligately Anaerobic Bacteria
               and Archaea . . . . . . . . . . . . . . . . . . . . . . . . . . . . . . . .   17
   3.2  CO-utilizing Sulfate-Reducing Bacteria and Archaea  . . . . . . . .   19
   3.3  CO-utilizing Sulfur-Reducing Archaea . . . . . . . . . . . . . . . . . .   22
   3.4  CO-utilizing Acetogenic Bacteria . . . . . . . . . . . . . . . . . . . . . .   22
   3.5  CO-utilizing Methanogenic Archaea . . . . . . . . . . . . . . . . . . . .   23
   3.6  CO-utilizing Iron-Reducing Bacteria and Archaea . . . . . . . . . . .   26

4  Biotechnological Applications of Thermophilic
   Carboxydotrophs . . . . . . . . . . . . . . . . . . . . . . . . . . . . . . . . . . .   29
   4.1  Electricity Production from CO/Syngas-fed Microbial
        Fuel Cell . . . . . . . . . . . . . . . . . . . . . . . . . . . . . . . . . . . . . .   31
        4.1.1  Overview of Microbial Fuel Cells (MFCs) . . . . . . . . . . .   31
        4.1.2  CO/Syngas as Substrates for Microbial Fuel
               Cells (MFCs) . . . . . . . . . . . . . . . . . . . . . . . . . . . . . .   38
        4.1.3  Thermophilic Carboxydotrophs in CO/Syngas-fed
               Microbial Fuel Cells (MFCs) . . . . . . . . . . . . . . . . . . .   41
        4.1.4  Design Considerations of MFCs Operating
               at Thermophilic Temperatures . . . . . . . . . . . . . . . . . . .   48
   4.2  Biofuel and Organic Acid Production by Thermophilic
        Carboxydotrophs from Synthesis Gas (Syngas) Fermentation . . . .   50
        4.2.1  Microbiology of Carboxydotrophic Microorganisms
               Capable of Converting Syngas to Biofuels and Organic
               Acids . . . . . . . . . . . . . . . . . . . . . . . . . . . . . . . . . . . .   57

  4.2.2   Biochemistry of Syngas Fermentation . . . . . . . . . . . . . .   67
  4.2.3   Products of Syngas Fermentation. . . . . . . . . . . . . . . . .   72
  4.2.4   Potential Improvements . . . . . . . . . . . . . . . . . . . . . . .   77
4.3   Bioremediation of Toxic Compounds with Thermophilic
      Carboxydotrophs . . . . . . . . . . . . . . . . . . . . . . . . . . . . . . . .   81
  4.3.1   Reductive Dehalogenation of Trichloroethylene
          by *Methanosarcina Thermophila* . . . . . . . . . . . . . . . . . .   83
  4.3.2   Reductive/Substitutive Dechlorination of
          Tetrachloromethane ($CCl_4$) by *Clostridium*
          *Thermoaceticum* and *Methanobacterium*
          *Thermoautotrophicum* . . . . . . . . . . . . . . . . . . . . . . . .   85
  4.3.3   Transformation of 2,4,6,-trinitrotoluene (TNT)
          by a Carboxydotrophic Sulfate-reducing Anaerobe. . . . . .   87
4.4   Thermophilic Carboxydotrophs as Biosensors for CO
      Detection . . . . . . . . . . . . . . . . . . . . . . . . . . . . . . . . . . . . .   88
  4.4.1   Mediators. . . . . . . . . . . . . . . . . . . . . . . . . . . . . . . . . . .   92
  4.4.2   Enzyme Electrodes . . . . . . . . . . . . . . . . . . . . . . . . . . . .   92
  4.4.3   Gas Biosensors . . . . . . . . . . . . . . . . . . . . . . . . . . . . . . .   93
  4.4.4   Detection of CO from the Environment . . . . . . . . . . . . .   94

5   Conclusions . . . . . . . . . . . . . . . . . . . . . . . . . . . . . . . . . . . . . . .   103

References. . . . . . . . . . . . . . . . . . . . . . . . . . . . . . . . . . . . . . . . . . . .   105

# Abstract

Carbon monoxide (CO) is a color- and odorless gas toxic to many organisms due to its high affinity to metal-containing enzymes. Despite its toxicity on the majority of living organisms on our planet, numerous microorganisms both aerobic and anaerobic can use CO as source of carbon and energy for growth. CO-oxidizing microorganisms or simply carboxydotrophs are group of microorganisms capable of using CO as their energy source. The oxidation of CO is coupled to acetogenesis, methanogenesis, hydrogenogenesis, sulfate/sulfur reduction, or iron reduction. Although as diverse as the organisms capable of it, any CO-dependent energy metabolism known depends on the presence of carbon monoxide dehydrogenase. CO metabolism is an important part of the global carbon cycle and understanding the processes involved will be helpful in developing or improving procedures employing CO for remediation or as a source of biomass or fuel. This review summarizes recent insights into the ecology, biochemistry, and applications of thermophilic carboxydotrophs in industrial processes and environmental biotechnology including the production of biofuels, the production of electricity through microbial fuel cell technology, bioremediation and biodegradation of toxic chemicals, and the detection of CO using the biosensor technology.

**Keywords** CO-oxidizing bacteria · Carbon monoxide dehydrogenase · Synthesis gas · Biofuel · Microbial fuel cell · Bioremediation · Biosensor · Hydrogenogens · Acetogens · Methanogens · Thermophiles · Carboxydotrophs

# Chapter 1
# Introduction

Carbon monoxide (CO) is an atmospheric trace gas. Its concentration in the Earth's atmosphere ranged between 0.06 and 0.15 ppm (IPPC 2001). CO is a potent electron donor ($E^{o'}-524$ to 558 mV for the $CO/CO_2$ couple) (Grahame and DeMoll 1995), and although it may be inhibitory to iron proteins of both aerobes and anaerobes (Adam 1990), CO is utilized by many microorganisms. CO-oxidizing microorganisms or simply carboxydotrophs are taxonomically diverse group of microorganisms and represent a type of chemolithoautotrophic metabolism capable of catalyzing the oxidation of CO to $CO_2$. The term carboxydotrophs was originally coined for microbes with aerobic, respiratory, chemolithoautotrophic utilization of CO as a sole source of carbon and energy (Meyer and Schlegel 1983). However, the term will be used here for all microbes capable of using CO as their energy source, because most CO-utilizing microorganisms studied to date are either facultative anaerobes or obligate anaerobes. These microorganisms are able (1) to catalyze the oxidation of CO to $CO_2$; (2) to use the electrons derived from this reaction for growth; (3) to assimilate parts of the $CO_2$ ribulose biphosphate pathway; and (4) to withstand CO inhibition. Certain fungi (Inman and Ingersoll 1971); algae (Chappelle 1962); actinomycetes and streptomycetes (Bartholomew and Alexander 1982; Gadkari et al. 1990); mycobacteria (Kim and Park 2014); ammonium oxidizers (Bedard and Knowles 1989; Jones and Morita 1983); methanotrophs (Bedard and Knowles 1989; Ferenci et al. 1975; Hubley et al. 1974); nitrogen-fixing bacteria (Chappelle 1962); photosynthetic bacteria (Pakpour et al. 2014); acetogens (Henstra et al. 2007a), sulfate-reducing bacteria (Henstra et al. 2007b); iron-reducing bacteria (Slepova et al. 2009); and hydrogenogenic bacteria and archaea (Sokolova et al. 2004a) are able to oxidize CO to $CO_2$. Besides CO, many carboxydotrophic microorganisms can grow with hydrogen plus carbon dioxide (Henstra et al. 2007b), indicating that their autotrophic capabilities are not restricted as those of non-carboxydotrophic, classical hydrogen-oxidizing bacteria. Most carboxydotrophs are facultative lithotrophs in that they are able to use a wide variety of organic substrates for heterotrophic growth (Drake et al. 2002; Sokolova et al. 2002, 2007; Henstra and Stams 2004; Bertoldo and

© The Author(s) 2014
S.M. Tiquia-Arashiro, *Thermophilic Carboxydotrophs and their Applications in Biotechnology*, Extremophilic Bacteria, DOI 10.1007/978-3-319-11873-4_1

Antranikian 2006; Balk et al. 2009; Sokolova and Lebedinsky 2013; Nguyen et al. 2013).

CO is an abundant atmospheric pollutant generated to a large extent by incomplete combustion of fossil fuels in domestic and industrial processes. For instance, blast furnace gas contains 25 % of CO and automobile exhaust gas contains 0.5–12 % of CO (Colby et al. 1985). CO occurs in the troposphere at a concentration of 0.1 ppm and in polluted urban areas; its concentration has been reported to reach levels of 50–100 ppm (Colby et al. 1985). Furthermore, CO contributes to ground-level ozone and indirect greenhouse warming (King 1999). Photolysis of dissolved organic matter generates CO in oceans and fresh waters (Conrad and Seiler 1980; Bullister et al. 1982; Conrad et al. 1982; Zuo and Jones 1995). CO is also liberated with volcanic gases at concentrations ranging between 1 and 2 % (Symonds et al. 1994). CO can be formed as a product of microbial metabolism by aerobic microorganisms during heme degradation (Ratliff et al. 2001; O'Brien et al. 1984). Production of low amounts of CO was observed in some thermophilic acetogenic bacteria and methanogenic archaea (Svetlichny et al. 1991a).

Thermophiles are probably one of the most interesting varieties of extremophilic microorganisms because they can thrive at temperatures over 50 °C. Among all thermophiles, there is much higher number of anaerobes than aerobes. This is perhaps due to the fact that oxygen is much less soluble at higher temperatures and therefore is not available to organisms in metabolic processes. Based on their optimal temperature, thermophiles can be subdivided into three main groups: moderate thermophiles, with an optimal temperature between 50 and 64 °C, and a maximum at 70 °C; extreme thermophiles, with an optimal temperature of between 65 and 80 °C; and finally hyperthermophiles, with an optimum temperature above 80 °C and a maximum above 90 °C (Stetter 1996). The main interest in thermophiles during the recent decades has been on issued dealing with basic and applied research. In addition, the discovery of many novel hyperthermophilic archaea, of which many can grow at 100 °C and above and a few even up to 121 °C, has attracted great interest among the scientific community. Thermophiles are certainly interesting in terms of biotechnologies, as many chemical industrial processes employing high temperatures have to be lowered in order to use bioprocess from mesophiles, and this could be avoided using enzymes from thermophiles (Wiegel and Ljungdahl 1986). Thermophiles are currently a source for a variety of high-value compounds such as antioxidants, biostabilizers, compatible solutes, biosurfactants, and biopolymers. Their unique enzymes are able to catalyze reactions under high temperature and can serve as excellent sources of thermostable biocatalysts. Research into thermophilic microorganisms has demonstrated that their thermotolerant proteins are generally more stable than other proteins and retain this property when cloned and expressed in mesophilic bacteria (Connaris et al. 1998; Hayakawa et al. 2009). Enzymes active and stable at elevated temperatures were investigated for biotechnological applications, particularly for bioconversion, biorefining, and biofuel production (Donaghy et al. 2000; Vielle and Zeikus 2001; Li et al. 2005; Razvi and Scholtz 2006; Turner et al. 2007; Liang et al. 2010;

Willies et al. 2010; Littlechild 2011). Further biotechnologies involving thermophiles cover the fields of biomass and complex organic molecule degradation (Blumer-Schuette et al. 2008; Suryawanshi et al. 2010; Sizova et al. 2011), metal leaching (Chen and Pan 2010), production of compatible solutes (Empadinhas and da Costa 2006), and water treatment technology (Liao et al. 2010). The potential application of thermophilic microorganisms for the production of biofuels, including methane and hydrogen as well as ethanol from biomass by means of thermophilic biological processes, was investigated over the past decades (Wiegel 1980; Henstra et al. 2007b; Koskinen et al. 2008; Shaw et al. 2008; Taylor et al. 2009; Kongjan et al. 2010; Roberts et al. 2010).

Recent advances in microbial physiology demonstrate that CO is a readily used substrate by thermophilic microorganisms and may be employed in novel biotechnological processes for the production of bulk and fine chemicals or in biological treatment of waste streams (Sipma et al. 2006). This brief focuses on the ecology, biochemistry, and applications of thermophilic carboxydotrophs in industrial processes and environmental biotechnology including the production of biofuels, the production of electricity through microbial fuel cell technology, bioremediation, and biodegradation of toxic chemicals, and the detection of CO using the biosensor technology.

# Chapter 2
# Microbial CO Metabolism

CO metabolism begins with a reaction that can be considered a thermodynamically favorable disproportionation, resulting in $CO_2$ and a pair of reducing equivalents, or molecular hydrogen as products:

$$CO + H_2O \rightarrow CO_2 + 2H^+ + 2e^-$$

The enzyme catalyzing this reaction is carbon monoxide dehydrogenase (CODH). This enzyme functions to either oxidize CO, synthesize acetyl-CoA, or cleave acetyl-CoA in a variety of energy-yielding pathways (Hausinger 1993). It contains iron (non-heme) and either molybdenum (in aerobes) or nickel (in anaerobes) in the active site (Ragsdale 2004). In aerobic carboxydotrophs, the CODH is monofunctional, which consist of a dimer of heterotrimers containing FAD and molybdopterin cytosine dinucleotide cofactors (Dobbek et al. 2002). The high-resolution crystal structures of the MoFeS CODHs from *Oligotropha carboxidovorans* (Gremer et al. 2000) or *Hydrogenophaga pseudoflava* (Hanzelmann et al. 2000) show a dimer of two heterotrimers in a $(LMS)_2$ subunit structure. Each heterotrimer is composed of a molybdenum protein (L subunit), a flavoprotein (M subunit), and an iron–sulfur protein (S subunit). The molybdoprotein carries the active site, which contains a 1:1 molar complex molybdopterin cytosine dinucleotide and a molybdenum atom. The iron–sulfur protein contains the type I and type II [2Fe-2S] centers. The flavoprotein contains the flavin adenine dinucleotide (FAD) cofactor and shows a new flavin-binding type (Gremer et al. 2000). The genes encoding the aerobic CODH are denoted by *cox* (carbon monoxide oxidase genes) (Hugendieck and Meyer 1992). Two forms of aerobic CODH have been identified so far: form I and form II. Form I has been specifically characterized for its ability to oxidize CO. Form II, which is phylogenetically close it, but distinct from form I, is a putative CODH, the true function of which remains uncertain. In contrast to form II CODH, the active site of form I CODH contains a unique catalytically essential loop of four amino acids, cysteine, serine, phenylalanine, and arginine, and a copper atom linked to the active site cysteine sulfur and to the molybdenum atom. In these bacteria, the

S.M. Tiquia-Arashiro, *Thermophilic Carboxydotrophs and their Applications in Biotechnology*, Extremophilic Bacteria, DOI 10.1007/978-3-319-11873-4_2

reducing equivalents resulting from CO oxidation are funneled through a CO-insensitive respiratory chain via ubiquinone and cytochromes ultimately resulting in oxygen reduction (Frunzke and Meyer 1990).

Aerobic CO oxidation can be viewed analogous to aerobic $H_2$ oxidation carried out by knallgas bacteria (Schlegel 1966); however, $H_2$ is not an intermediate in CO-dependent $O_2$ reduction (Meyer and Schlegel 1983). In some aerobic carboxydotrophs, energy conserved from CO metabolism can be used to fix $CO_2$ to biomass. This process typically involves the Calvin–Benson–Bassham (CBB) cycle, which is based on the enzyme ribulose-1,5-bisphosphate carboxylase/oxygenase (Ragsdale 2004). Other aerobic carboxydotrophs seem unable to fix $CO_2$ by the CBB cycle or other mechanisms. In these organisms, CO can provide a supplemental energy source without contributing directly to biomass.

Anaerobic carboxydotrophs have CODHs that differ distinctly from that of aerobic carboxydotrophs, in part because it contains nickel instead of molybdenum as a metal cofactor. The NiFeS CODHs are either monofunctional or bifunctional (Ferry 1995). They form complex with acetyl coenzyme A (acetyl-CoA) synthase (ACS). The monofunctional CODH from phototrophic bacterium *Rhodospirillum rubrum* (Bonam and Ludden 1987) is inducible in the dark under anaerobic conditions in the presence of CO, shows a micromolar $K_m$ for CO (Bonam and Ludden 1987), and contains a proposed nickel–iron–sulfur cluster (cluster C) (Heo et al. 1999) and a conventional [4Fe–4S] cluster (cluster B) (Hu et al. 1996). Radiolabeling studies suggested a catalytically essential non-substrate CO ligand ($CO_L$) to the Fe atom in putative [Fe–Ni] center cluster C (Heo et al. 2000). Acetogens and methanogens employ the bifunctional CODH–ACS (Ferry 1995). The enzymes are tetramers of two different subunits or pentamers of five different subunits. The subunits harboring the CODH activity contain cluster C and cluster B (Ferry 1995), which is similar to the function of *R. rubrum* CODH. Phototrophic anaerobic carboxydotrophs grow by converting CO to $H_2$, a process initiating with CODH that contains nickel and iron–sulfur centers (Drennan 1991). Acetogenic carboxydotrophs employ nickel/iron–sulfur CODH to synthesize acetyl-CoA from a methyl group, CO, and CoA (Ragsdale and Kumar 1996). A similar enzyme is responsible for the cleavage of acetyl-CoA by anaerobic *Archaea* that obtain energy by fermenting acetate to $CH_4$ and $CO_2$. Sulfate-reducing carboxydotrophs from *bacteria* and *archaea* also utilize CODH to cleave acetyl-CoA yielding methyl and carbonyl groups. These microbes obtain energy for growth via a respiratory pathway in which the methyl and carbonyl groups are oxidized to $CO_2$, and the sulfate is reduced to sulfide (Ferry 1995). CO is considered an excellent source of energy since the redox potential ($E^{o'}$) of the $CO_2$/CO couple is very low (−524 to 558 mV) (Grahame and DeMoll 1995). However, relatively few anaerobic microbes capable of utilizing CO as their sole source of energy have been described. Known respiratory processes, which are coupled to anaerobic CO oxidation, are illustrated in Fig. 2.1. These respiratory processes include proton respiration (hydrogenogenesis) (Henstra et al. 2007b), sulfate (or sulfur) respiration (desulfurication) (Rabus et al. 2006), and carbonate respiration (acetogenesis and methanogenesis) (Drake et al. 2002).

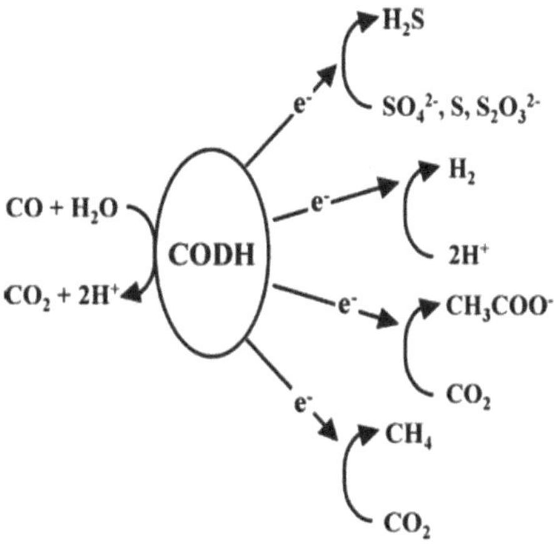

**Fig. 2.1** Scheme of anaerobic respirations, which can be coupled to CO oxidation. CODH carbon monoxide dehydrogenase; $-e^- \rightarrow$ electron transport chain, some components of which are unknown (Oelgeschlager and Rother 2008)

Anaerobic carboxydotrophic hydrogenogenic microorganisms conserve energy by oxidation of CO to $CO_2$ coupled to reduction of protons to $H_2$. These reactions are catalyzed by CO dehydrogenase and hydrogenase, respectively. These two enzymes must conserve energy in a yet-unknown energy-conserving mechanism as they do not fit in classical substrate-level phosphorylation (SLP) and electron transfer phosphorylation (ETP) theories (Hedderich 2004). Hydrogenases catalyze the reduction of protons to $H_2$ or the reverse reaction according to:

$$2H^+ + 2e^- \leftrightarrow H_2$$

The physiological function of hydrogenases is generally restricted to one of these directions and is referred to as hydrogen-uptake hydrogenase or hydrogen-evolving hydrogenase. Recently, the classification and phylogeny of hydrogenases were reviewed by Vignais et al. (2001). Three classes of hydrogenases are recognized based on phylogeny and metal content (also transition-metal content or $H_2$-activating site content). The first and largest class is formed by the [NiFe]-hydrogenases. The second class, [FeFe]-hydrogenases only, contains Fe in their active site, while the third class is formed by hydrogenases that until recently were named "metal-free" hydrogenases (Berkessel and Thauer 1995). The latter class was discovered in methanogens, where they catalyze the reduction of $F_{420}$ with $H_2$ in complex with methylenetetrahydromethanopterin dehydrogenase under nickel-deprived conditions (Zirngibl et al. 1992; Afting et al. 1998, 2000). Now they are referred to as iron–sulfur cluster-free hydrogenases, since the presence of a novel light-sensitive iron-coordinating cluster in this type of hydrogenases was found (Buurman et al. 2000; Lyon et al. 2004; Shima et al. 2004).

Awareness that a unique subclass of [NiFe]-hydrogenases exists has grown in the past decade. These hydrogenases were indicated as energy-converting hydrogenases (ECH), after their capacity to couple proton translocation to the reduction of protons or oxidation of molecular $H_2$, and were recently reviewed by Hedderich et al. (2004). A limited sequence similarity and a deviating enzyme topology mark the main differences between ECH and other [NiFe]-hydrogenases. These ECH are membrane-bound enzyme complexes and play a key role in energy generation in the carboxydotrophic hydrogenogenic metabolism (Hedderich et al. 2004). The ECH couples the oxidation of $H_2$ or reduction of protons to translocation of protons over the cytoplasmic membrane. The electrochemical gradient of protons over the membrane is generally referred to as proton motive force (pmf) and is the driving force for ATP synthesis. Translocation of protons by ECH may generate the pmf depending on its direction. Reduction of protons to form $H_2$ is coupled to the generation of a pmf, which in turn drives ATP synthesis. The number of suitable electron donors for reduction of the protons is limited by the relatively low electrode potential of the $H^+/H_2$ couple ($E^{\circ\prime}$ −414 mV). The $CO_2/CO$ couple ($E^{\circ\prime}$ −520 mV) is sufficiently low to drive proton translocation by ECH, but also formate or reduced ferredoxin generated by pyruvate:ferredoxin oxidoreductase in fermentative metabolisms serve as electron donors for ECH (Bagramyan and Trchounian 2003; Sapra et al. 2003; Hedderich et al. 2004; Soboh et al. 2004). Reduced ferredoxin may donate electrons directly to ECH while CODH and formate dehydrogenase form a complex with ECH. In these complexes, a ferredoxin-like subunit facilitates electron transfer. In the carboxydotrophic hydrogenogenic metabolism, ECH together with CODH play an important role as described for *Carboxydothermus hydrogenofor-mans* and *R. rubrum* (Hedderich et al. 2004; Fox et al. 1996a,b). These organisms contain similar enzymatic systems that catalyze the conversion of CO into $H_2$. Genes that code for the involved enzymes are arranged in two gene clusters that share high sequence similarity between both organisms. One cluster comprises the genes for ECH, the other cluster encodes for a CODH and *CooF*. One functional CO-oxidizing $H_2$-evolving complex is formed with these subunits, as was shown for *C. hydro-genoformans* (Soboh et al. 2002). Coupling of proton translocation to oxidation of CO by *C. hydrogenoformans* and *R. rubrum* enables them to use CO as sole source of energy and to grow with the formation of $H_2$.

The possible role of CO as electron donor in anaerobic respiration has received little attention, and the number of species known to use CO in anaerobic respiration is still limited. *Moorella thermoacetica* can grow chemolithotrophically with CO as electron donor and nitrate as electron acceptor (Drake and Daniel 2004; Frostl et al. 1996). Furthermore, *Thermosinus carboxydivorans* reduces ferric iron and selenite with CO as electron donor (Sokolova et al. 2004b), and *C. hydrogenoformans* reduces fumarate and 9,10-anthraquinone-2,6-disulfonate (AQDS) with CO as electron donor (Henstra and Stams 2004). *C. hydrogenoformans reduces* nitrate, thiosulfate, sulfur, and sulfite with lactate as electron donor, but according to Henstra and Stams (2004), CO might be able to serve as electron donor as well. However, *T. carboxydivorans* and *C. hydrogenoformans* form $H_2$ from CO, which might be the actual electron donor for these reductions. In contrast, *Thermoterra-bacterium ferrireducens* does not form hydrogen with CO, but is able to reduce

AQDS and fumarate with CO. Besides AQDS, *T. ferrireducens* may reduce Fe (III), nitrate, sulfite, thiosulfate, and sulfur with CO as well, as it does with $H_2$ or lactate (Henstra and Stams 2004; Slobodkin et al. 1997).

The exact range of microorganisms capable to use CO is still unclear, as CO utilization is rarely tested in growth studies. Although CO may initially inhibit growth, adaptation to CO can occur after long-term incubation or multiple transfers with increasing CO levels (Rother and Metcalf 2004; O'Brien et al. 1984). Furthermore, growth on CO may require different nutrients or concentrations (Kerby et al. 1995). The ability of CO oxidation to $CO_2$ seems ubiquitously present in nature and has an ancient origin, as mentioned by Ferry (1995) and Hedderich (2004).

# Chapter 3
# CO-oxidizing Microorganisms

CO is metabolized by a wide variety of microorganisms. A sharp division exists between aerobic and anaerobic species, as they contain fundamentally different enzyme systems for CO biotransformation. Aerobic CO-oxidizing bacteria may be divided in two groups: metabolic, in which CO oxidation provides energy for growth, and co-metabolic, in which CO is used as pseudo-substrate for the enzyme system, but does not provide a nutritional value (Colby et al. 1985). The latter is observed during aerobic CO oxidation by methane-oxidizing bacteria employing the methane monooxygenase complex, which is rather unspecific with respect to its substrate (Higgins et al. 1980; Daniels et al. 1977). Aerobic metabolic CO-oxidizing bacteria or aerobic carboxydotrophs use CO as a source of energy, which is oxidized with $O_2$ as terminal electron acceptor. These bacteria contain a specific CO-tolerant cytochrome b1 oxidase and $O_2$-insensitive Mo–Fe–flavin carbon monoxide dehydrogenase. The diversity and ecology of bacteria that grow aerobically with CO has been intensively studied and reviewed (Zavarzin and Nozhevnikova 1977; Meyer et al. 1990; Conrad 1996; King and Webber 2007). Aerobic carboxydotrophs include CO-utilizing microorganisms that belong to α-*Proteobacteria, Firmicutes,* and *Actinobacteria* (Table 3.1).

Aerobic CO oxidation merely results in the production of $CO_2$ and biomass in case of an energy-yielding CO metabolism. Anaerobic conversion of CO, on the other hand, results in the production of a range of other compounds. Representatives of various groups, e.g., homoacetogens, methanogens, and Sulfate-reducing bacteria, have been identified in CO metabolism. An overview of anaerobic microorganisms known to metabolize CO as sole source of carbon and energy is presented in Table 3.2. Many microorganisms that use carbon monoxide as an energy source are found in high-temperature environments such as geothermal areas. Researchers think that these carboxydotrophs may be involved in reducing potentially toxic carbon monoxide hot spots by combine with water to form hydrogen, carbon dioxide, and acetate, which are in turn used for thermophilic energy conservation and carbon sequestration mechanisms. Natural and anthropogenic hot environments contain appreciable levels of carbon monoxide. The anaerobic microbial communities play important role in the conversion of CO in these environments. Thermophilic carboxydotrophic anaerobes include

© The Author(s) 2014
S.M. Tiquia-Arashiro, *Thermophilic Carboxydotrophs and their Applications in Biotechnology*, Extremophilic Bacteria, DOI 10.1007/978-3-319-11873-4_3

**Table 3.1** Representative aerobic microorganisms that utilize CO as a sole C and energy source

| CO-utilizing aerobic microorganisms | Isolation source | Reference |
|---|---|---|
| *α-Proteobacteria* | | |
| *Oligotropha carboxidovorans* | Wastewater | Meyer and Schlegel (1983) |
| *Pseudomonas thermocarboxydovorans* | Compost | Lyons et al. (1984) |
| *Pseudomonas carboxydohyrogena* | Sewage | Meyer and Schlegel (1983) |
| *Bradyrhizobium japonicum* USDA 110 | Rhizosphere | Lorite et al. (2000), Keneko et al. (1990) |
| *Firmicutes* | | |
| *Bacillus schlegelii* | Freshwater sediment | Meyer and Schlegel (1983), Krueger and Meyer (1984) |
| *Actinobacteria* | | |
| *Streptomyces thermoautotrophicus* | Coal heap | Gadkari et al. (1990) |
| *Mycobacterium smegmatis* | Environment, human | King (2003), Park et al. (2003) |
| *Mycobacterium gordonae* | Human, water | King (2003) |
| *Mycobacterium tuberculosis* H37Ra | Lung isolate | King (2003), Park et al. (2003) |
| *Mycobacterium* sp. JC1 | Soil | Park et al. (2003), Kang and Kim (1999) |

methanogens, acetogens, hydrogens, sulfate reducers, sulfur reducers, and ferric iron reducers (Table 3.2). Among these groups of carboxydotrophs, hydrogenogens are the most numerous and based on available data and they are most important in CO biotransformation in hot environments (Sokolova et al. 2009; Sokolova and Lebedinsky 2013).

## 3.1 CO-utilizing Hydrogenogenic Bacteria and Archaea

Hydrogenogens can grow by oxidizing CO and subsequently reducing the protons derived from $H_2O$ to form molecular hydrogen:

$$CO + H_2O \rightarrow CO_2 + H_2 \quad \Delta G° = -20\,\text{kJ/mol CO}$$

This process proceeds in the absence of an electron acceptor and is analogous to the water–gas–shift reaction used to deplete gas mixtures of CO (Graven and Long 1954). Hydrogenogens belonging to the domain *Bacteria* and *Archaea* have been isolated from various locales around the world (Table 3.2). Currently, this carboxydotrophic hydrogenogenic metabolism is found in three distinct groups of

**Table 3.2** Representative anaerobic carboxydotrophs isolated from different environments

| Organism | Isolation source | Optimal temperature (°C) | Reference |
|---|---|---|---|
| **Hydrogenogens** | | | |
| *Caldanaerobacter subterraneus sp. pacificus* | Mud | 70 | Fardeau et al. (2004), Sokolova et al. (2001) |
| *Carboxydocella thermautotrophica* | Cyanobacterial mat and mud | 58 | Sokolova et al. (2002) |
| *Carboxydocella sporoproducens* | Cyanobacterial mat and mud | 60 | Slepova et al. (2006) |
| *Carboxythermus hydrogenoformans* | Mud from hot swamp | 70–72 | Svetlichny et al. (1991b) |
| *Carboxydothermus siderophilus* | Pink filaments | 72 | Slepova et al. (2009) |
| *Dictyoglomus carboxydivorans* | Water and mud | 80 | Kochetkova et al. (2011) |
| *Thermincola carboxydiphila* | Mud and cyano-bacterial mat | 51–72 | Sokolova et al. (2005) |
| *Thermincola ferriacetica* | Water and ocher deposits | 65 | Zavarzina et al. (2007) |
| *Thermolithobacter carboxydivorans* | Water and sediment | 80 | Sokolova et al. (2007) |
| *Thermosinus carboxydivorans* | Mud and water | 60 | Sokolova et al. (2004b) |
| *Thermococcus* AM4 | Deep-sea hydro-thermal vent | 82 | Sokolova et al. (2004a) |
| *Thermococcus onnurineus* | Hydrothermal vent | 80 | Bae et al. (2006), Lee et al. (2008) |
| *Thermofilum carboxyditrophus* | Water and mud | 90 | Kochetkova et al. (2011) |
| **Sulfate-reducing bacteria** | | | |
| *Archaeoglobus fulgidus* | Hot oil field waters | 76 | Stetter (1988), Henstra et al. (2007a) |
| *Desulfotomaculum carboxydivorans* | Sludge from an anaerobic bioreactor | 55 | Parshina et al. (2005a) |
| *Desulfotomaculum kuznetsovii* | Petroleum refinery | 60–65 | Parshina et al. (2005a) |
| *Desulfomonile tiedjei* | Sewage sludge | 37 | DeWeerd et al. (1990) |
| *Desulfovibrio vulgaris (strain Madison)* | Soil | 37 | Lupton et al. (1984) |

(continued)

**Table 3.2** (continued)

| Organism | Isolation source | Optimal temperature (°C) | Reference |
|---|---|---|---|
| *Desulfovibrio desulfuricans* | Mud, soil, rumen sheep | 37 | Davidova et al. (1994) |
| *Desulfosporosinus orientis* | Soil | 35 | Klemps et al. (1985) |
| *Desulfotomaculum nigrificans* | Soil | 35 | Klemps et al. (1985) |
| *Desulfotomaculum thermo-benzoicum subsp. thermosyntrophicum* | Anaerobic sludge bed reactor | 55 | Parshina et al. (2005a) |
| **Acetogens** | | | |
| *Moorella thermoacetica* | Horse feces | 55 | Daniel et al. (1990) |
| *Moorella thermoautotrophica* | Mud and wet soils | 58 | Savage et al. (1987) |
| *Clostridium autoethanogenum* | Rabbit feces | 37 | Abrini et al. (1994) |
| *Oxobacter pfennigii* | Rumen of cattle | 36–38 | Krumholz and Bryant (1985) |
| *Clostridium ljungdahlii* | Chicken yard waste | 37 | Tanner et al. (1993) |
| *Peptostreptococcus productus* | Sewage sludge digester | 37 | Lorowitz and Bryant (1984) |
| *Acetobacterium woodii* | Mud | 30 | Sharak Genthner and Bryant (1987) |
| *Eubacterium limosum* | Rumen of sheep | 38–39 | Sharak Genthner and Bryant (1982) |
| *Butyribacterium methylotrophicum* | Sewage sludge digester | 37 | Grehtlein et al. (1991) |
| **Ferric iron reducer** | | | |
| *Carboxydothermus pertinax* | Volcanic acidic hot spring | 65 | Yoneda et al. (2012) |
| *Carboxydothermus siderophilus* | Kamchatka hot spring | 65 | Slepova et al. (2009) |
| *Thermosinus carboxydivorans* | Mud and water | 60 | Sokolova et al. (2004b) |
| *Thermincola ferriacetica* | Terrestrial hydrothermal spring | 57–60 | Zavarzina et al. (2007) |
| **Methanogens** | | | |
| *Methanobacterium thermoautotrophicum* | Formation water of oil-bearing rocks | 65 | Daniels et al. (1977) |
| *Methanosarcina acetivorans* | Marine mud | 37 | Moran et al. (2008) |
| *Methanosarcina barkeri* | Anaerobic sewage digester | 37 | O'Brien et al. (1984) |

(continued)

**Table 3.2**  (continued)

| Organism | Isolation source | Optimal temperature (°C) | Reference |
|----------|------------------|--------------------------|-----------|
| *Methanosarcina thermophila* | Anaerobic sludge digester | 50 | Zinder et al. (1984, 1979) |
| *Methanothermobacter thermoautotrophicus* | Municipal waste-treatment facility | 65–70 | Daniels et al. (1977) |
| **Sulfur reducer** | | | |
| *Thermoproteus tenax* | Solfatara mud hole | 88 | Fischer et al. (1983) |

prokaryotes, i.e., mesophilic Gram-negative bacteria, thermophilic Gram-positive bacteria, and thermophilic archaea (Table 3.3). The mesophilic Gram-negative bacteria are facultatively anaerobic, while the thermophilic Gram-positive bacteria are obligately anaerobic. Well-known hydrogenogens are *Rubrivivax gelatinosus* (formerly *R. gelatinosus*) (Dashekvicz and Uffen 1979), *Rhosodspirullum rubrum* (Kerby et al. 1995), and *Carboxydothermus hydrogenoformans* (Svetlichny et al. 1991b). Both photosynthetic bacteria, *Rhodospirillum rubrum* and *R. gelatinosus*, are able to grow anaerobically in the dark by converting CO to $CO_2$ and $H_2$ (Dashekvicz and Uffen 1979; Kerby et al. 1995).

## 3.1.1  CO-utilizing Facultatively Anaerobic Bacteria

Facultatively anaerobic bacteria that oxidize CO and evolve $H_2$ isolated thus far are Gram-negative mesophilic bacteria. Hydrogen is only produced under anaerobic conditions upon CO oxidation. Generally, growth rates on CO are low and high levels of CO are inhibitory. Non-sulfur purple bacteria form the predominant part of this group of bacteria. Oxidation of CO coupled to formation of equimolar amounts of $H_2$ was first discovered with *Rhodopseudomonas gelatinosa* (Dashekvicz and Uffen 1979; Uffen 1976), later reclassified as *Rhodocyclus gelatinosa* (Imhoff et al. 1984) and more recently as *R. gelatinosus* (Willems et al. 1991). Hydrogenogenic CO conversion was also observed with protein extracts of *R. rubrum* S1 (Uffen 1981), although the growth rates of *R. rubrum* S1 were considered too low for elucidating the microbiology of CO metabolism. However, Kerby et al. (1995) demonstrated that *R. rubrum* in fact was capable of a rapid anaerobic growth on CO in the dark, but only after increasing the nickel concentration in the medium. Two different strains of *R. rubrum* revealed different $NiCl_2$ requirements with concentrations exceeding 75 and 600 µM necessary for growth. Besides different nutrient requirements for growth on CO, the presence of $CO_2$ may have a stimulatory effect on growth and CO conversion, as recently reported for the homoacetogen *Moorella thermoacetica* (Drake and Daniel 2004). *R. rubrum* so far is the most studied

**Table 3.3** Representative hydrogenogenic CO-utilizing microorganisms

| Co-utilizing microorganisms | Optimum temperature (°C) | Optimum pH | Doubling time (h) | Reference |
|---|---|---|---|---|
| *Bacteria* | | | | |
| **Facultative anaerobes** | | | | |
| *Rubrivivax gelatinosa* | 34 | 6.8–6.9 | 6.7 | Uffen (1976), Dashekvicz and Uffen (1979) |
| *Rhodopseudomonas palustris* P4 | 30 | nd | 2 | Jung et al. (1999a) |
| *Rhodospirillum rubrum* | 30 | 6.8 | 8.4 | Kerby et al. (1995) |
| *Citrobacter* sp Y19 | 30–40 | 5.5–7.5 | 8.3 | Jung et al. (1999b) |
| **Obligate anaerobes** | | | | |
| *Carboxydothermus hydrogenoformans* | 70–72 | 6.8–7.0 | 2 | Svetlichny et al. (1991b) |
| *Carboxydothermus restrictus* | 70 | 7.0 | 0.3 | Svetlichny et al. (1994) |
| *Caldanaerobacter subterraneus* subsp. *pacificus* | 70 | 6.8–7.1 | 7.1 | Sokolova et al. (2001) |
| *Carboxydocella thermoautotrophica* | 58 | 7.0 | 1.1 | Sokolova et al. (2002) |
| *Thermosinus carboxydivorans* | 60 | 6.8–7.0 | 1.2 | Sokolova et al. (2004b) |
| **Archaea** | | | | |
| *Thermococcus* strain AM4 | 82 | 6.8 | nd | Sokolova et al. (2004a) |

*nd* not determined

organism, especially with respect to the properties of its CO dehydrogenase. A major drawback of photosynthetic bacteria in $H_2$ production from synthesis gas derived CO is the light requirement for optimal cell growth. *Rhodopseudomonas palustris* P4 (Jung et al. 1999a) was found capable of hydrogenogenic CO conversion when incubated anaerobically in the dark, although growth completely ceased in the absence of light. It was postulated that *R. palustris* P4 only obtains maintenance energy from the hydrogenogenic CO conversion (Jung et al. 1999a). Besides these phototrophic strains, only one non-phototrophic Gram-negative facultative anaerobe was described so far capable to convert CO to $H_2$, viz. *Citrobacter* strain Y19, isolated from an activated sludge plant (Jung et al. 1999b). However, as the growth of this organism under anaerobic conditions compared to aerobic conditions is low, a two-step cultivation of the biomass was proposed, i.e., an aerobic growth phase followed by an anaerobic CO conversion phase (Jung et al. 1999b). Nevertheless, separation of growth and bioconversion complicates the reactor operation considerably.

## 3.1.2 CO-utilizing Obligately Anaerobic Bacteria and Archaea

A rapidly increasing group of carboxydotrophic hydrogenogenic prokaryotes is formed by strict anaerobic thermophiles (Table 3.3). Conversion of CO to $H_2$ at elevated temperatures has been observed in freshwater as well as marine environments with temperatures ranging from 40 to 85 °C and pH between 5.5 and 8.5 (Osmolovskaya et al. 1999; Svetlichny et al. 1991b).

The representatives of thermophilic hydrogenic carboxydotrophic bacteria were first discovered by Svelichny et al. (1991) in hot springs of Kunashir Island. Currently, there are 17 hydrogenogenic carboxydotrophic species belonging to the genera *Carboxydothermus, Thermincola, Carboxydocella, Thermosinus, Caldanerobacter, Thermoanaerobacter, Thermolithobacter, Dictyoglomus, Thermococcus,* and *Thermophilum* (Sokolova and Lebedinsky 2013). All *Carboxythermus* species are obligate anaerobes, extreme thermophiles, and neutrophils. The optimum temperature for growth ranges between 65 and 70 °C. All *Carboxydothermus* can utilize CO, but they are facultatively carboxydotrophs and can grow chemolithotrophically on CO or on other substrates. The genus *Carboxydothermus* currently contains four species including *C. hydrogenoformans* (Svelichny et al. 1991), *C. siderophilus* (Slepova et al. 2009), *C. islandicus* (Novikov et al. 2011), and one non-hydrogenogenic species *C. ferrireducens* (Slobodkin et al. 2006). Currently, *C. hydrogenoformans* is the best-studied thermophilic carboxydotrophic hydrogenogen. Its complete genome was sequenced by Wu et al. (2005a). *C. hydrogenoformans* employs a pathway very similar to that of *R. rubrum* (Svetlichny et al. 1991b). However recently, it was shown that *C. hydrogenoformans* can grow nonhydrogenogenically with various electron donors such as formate, lactate, and glycerol, as well as electron acceptors such as 9,10-anthraquinone-2,6-disulfonate, sulfite, thiosulfate, sulfur, nitrate and fumarate (Henstra and Stams 2004).

The genus *Thermincola* contains three described species: *Thermincola carboxydiphila* (Sokolova et al. 2005), *T. ferriacetica* (Zavarzina et al. 2007), and *Thermincola potens* (Byrne-Bailey et al. 2010). They are strictly anaerobic and moderately thermophilic. *T. carboxydiphila* and *T. ferriacetica* grow on 100 % CO hydrogenogenically. During chemolithotrophic growth on CO, these species require additional sources such as acetate and yeast extract. *T. carboxydiphila* is an obligate carboxydotroph, while *T. ferriacetica* is a facultative carboxydotroph in that it can grow in the expense of ferric iron reduction with hydrogen or acetate or some other organic sources as carbon source.

The genus *Carboxydocella* contains four species. Three of which are thermophilic hydrogenogenic carboxydotrophic species, *Carboxydocella thermautotrophica* (Sokolova et al. 2002), *C. sporoproducens* (Slepova et al. 2006), and *C. ferrireducens* (Sokolova et al. 2009), and one non-carboxydotrophic species, *C. manganica* (Slobodkina et al. 2012). *Carboxydocella* species are strictly anaerobic, moderately thermophilic, and neutrophilic. All *Carboxydocella* species can grow at

100 % with the exception of *C. manganica* (Slobodkina et al. 2012). Among the three carboxydotrophic species, *C. thermautotrophica* is the only obligate carboxydotroph. The other two species are facultative carboxydotrophs and can be organotrophic on several substrates.

The genus *Thermosinus* is represented by *T. carboxydivorans* (Sokolova et al. 2004b). *T. carboxydivorans* grows organotrophically on some carbohydrates or on pyruvate. During growth on CO, sucrose, or lactose, it reduces ferric iron. It does not utilize lactate, acetate, formate, and $H_2$ either in the absence or in the presence of ferric iron, thiosulfate, sulfate, sulfite, elemental sulfur, or nitrate.

The genus *Caldanaaerobacter* includes *Caldanaaerobacter subterraneus* and *C. subterraneus* subsp. *pacificus* (Sokolova et al. 2001; Fardeau et al. 2004). Both strains are strict anaerobes, extreme thermophiles, and neutrophiles. In addition to lithotrophic hydrogenogenic growth on CO, they grow organotrophically on some fermentable substrates.

*Thermoanaerobacter* species are strictly anaerobic, thermophilic (growing between 55 and 75 °C), and most form spores. The only hydrogenogenic representative is *T. hydrosulfuricus* subsp. *carboxydivorans* (Balk et al. 2009). It can grow both in the presence and absence of thiosulfate on a number of fermentable substrates including peptone, various sugars, xylan, starch, pectin, inulin, and cellobiose. The main products of sugar fermentation are lactate, acetate, ethanol, alanine, $H_2$ and $CO_2$. In addition to thiosulfate, elemental sulfur, sodium sulfite, ferric iron, $MnO_2$, anthraquinone-2,6-disulfonate, and arsenate can serve as electron acceptors.

The single hydrogenogenic carboxydotrophic strain of the genus *Thermolithobacter* is T. carboxydivorans (Sokolova et al. 2007). It grows on CO hydrogenogenically. It does not grow on CO in the presence of $NO_3^-$, Fe(III) oxide/hydroxide, Fe(III) citrate, or $SO_3^{2-}$. It grows on CO in the presence of $SO_4^{2-}$, $S_2O_3^{2-}$, or fumarate, but the presence of electron acceptors does not stimulate growth. It does not reduce $SO_4^{2-}$, $S_2O_3^{2-}$, or elemental sulfur in media supplemented with yeast extract, formate, acetate, pyruvate, citrate, succinate, lactate, or $H_2:CO_2$. It also does not grow in $H_2:CO_2$ mixture with $NO_3^-$, $SO_3^{2-}$, or fumarate.

*Dictyoglomus carboxydivorans* is currently the single thermophilic hydrogenogenic bacterium (Kochetkova et al. 2011). The organism is an extremely thermophilic anaerobic bacterium, which grows on CO at a concentration not exceeding 15 % in the gas phase. It grows on CO significantly slower than other thermophilic hydrogenogens.

Members of the genus *Thermococcus* (hyperthermophilic euyarchaeotes) can grow by fermenting carbohydrates or peptides, or by employing a respiratory energy metabolism with sulfur as an electron acceptor (Bertoldo and Antranikian 2006). *Thermococcus* sp AM4 is a representative of the genus *Thermococcus* (Sokolova et al. 2004b). This strain was described as the first hydrogenogenic carboxydotrophic representative of the domain *Archaea*. It was isolated from a deep-sea hydrothermal vent. It is growing by CO oxidation coupled with $H_2$ production. Since CODH is usually not found among members of the *Thermococcales*, it was concluded that strain AM4 has acquired the trait by a recent lateral gene transfer event (Sokolova et al. 2004b). Strain SM4 grows at 60–90 °C and has a broad spectrum temperature interval of 70–80 °C. It grows on 100 % CO in the gas phase. In the absence of CO, strain AM4 grows on some peptide substrates with elemental sulfur as electron acceptor. Another hydrogenogenic species of the genus *Thermococcus* is *T. onnurineus* NA1 (Bae et al. 2006; Lee et al. 2008). Analysis of the whole-genome sequence of *T. onnurineus* revealed the presence of a gene cluster containing genes for carbon monoxide dehydrogenase and the energy-converting hydrogenase (Lee et al. 2008). This strain can grow hydrogenogenically in 100 % CO in the gas phase (Lee et al. 2008).

*Thermofilum carboxytrophus* strain 1505 is the representative of the genus *Thermofilum*. This organism grows chemolithotrophically on 45 % Co in the gas phase at 92 °C producing equimolecular quantities of $H_2$ and $CO_2$ (Kochetkova et al. 2011). Yeast extract at the concentration of 0.2 g $L^{-1}$ is required for growth.

Although carboxydotrophic hydrogenogens share similarities in growth physiology, they are phylogenetically divergent (Fig. 3.1), suggesting that anaerobic CO oxidation by was propagated frequently through horizontal gene transfer. For instance, the similarity of the CODH operon of the carboxydotrophic hydrogenogens in Fig. 3.1 disagrees with the divergence of the 16S rRNA sequences of these taxa in Fig. 3.2.

## 3.2 CO-utilizing Sulfate-Reducing Bacteria and Archaea

Members of the desulfuricants (sulfate reducers) also use CO as a source of energy (Sipma et al. 2006). The oxidation reaction proceeds as follow:

$$4CO + SO_4^{-2} + H^+ \rightarrow 4CO_2 + HS^- \ \Delta G° = -57.8\,\mathrm{kJ/mol\ CO}$$

Most sulfate-reducing bacteria reported to utilize CO as an energy source and convert it to $CO_2$ and $H_2$. Subsequently, $H_2$ is used for sulfate reduction (Rabus et al. 2006). Since CO-utilizing sulfate reducers are sensitive to elevated levels of CO and can tolerate only a few percent CO in the gas phase (Sipma et al. 2006), $H_2$ production from CO oxidation probably serves as a CO-detoxification pathway. One exception is *Desulfotomaculum carboxydivorans*, which not only grows in the presence of 100 % CO but is also capable of growing in the absence of sulfate as hydrogenogen (Parshina et al. 2005a). Some sulfate reducers including, *Desulfotomaculum thermobenzoicum* and *Desulfotomaculum kuznetsovii,* do not only produce $H_2$ as a transient intermediate, but also generate acetate from CO during growth (Parshina et al. 2005b).

One of the best-studied archaeal sulfate reducers is *Archaeoglobus fulgidus* (Stetter 1988). It has a growth optimum of 83 °C and oxidizes lactate to $CO_2$ with concominant reduction of sulfate to sulfide. For sulfate reduction, the same

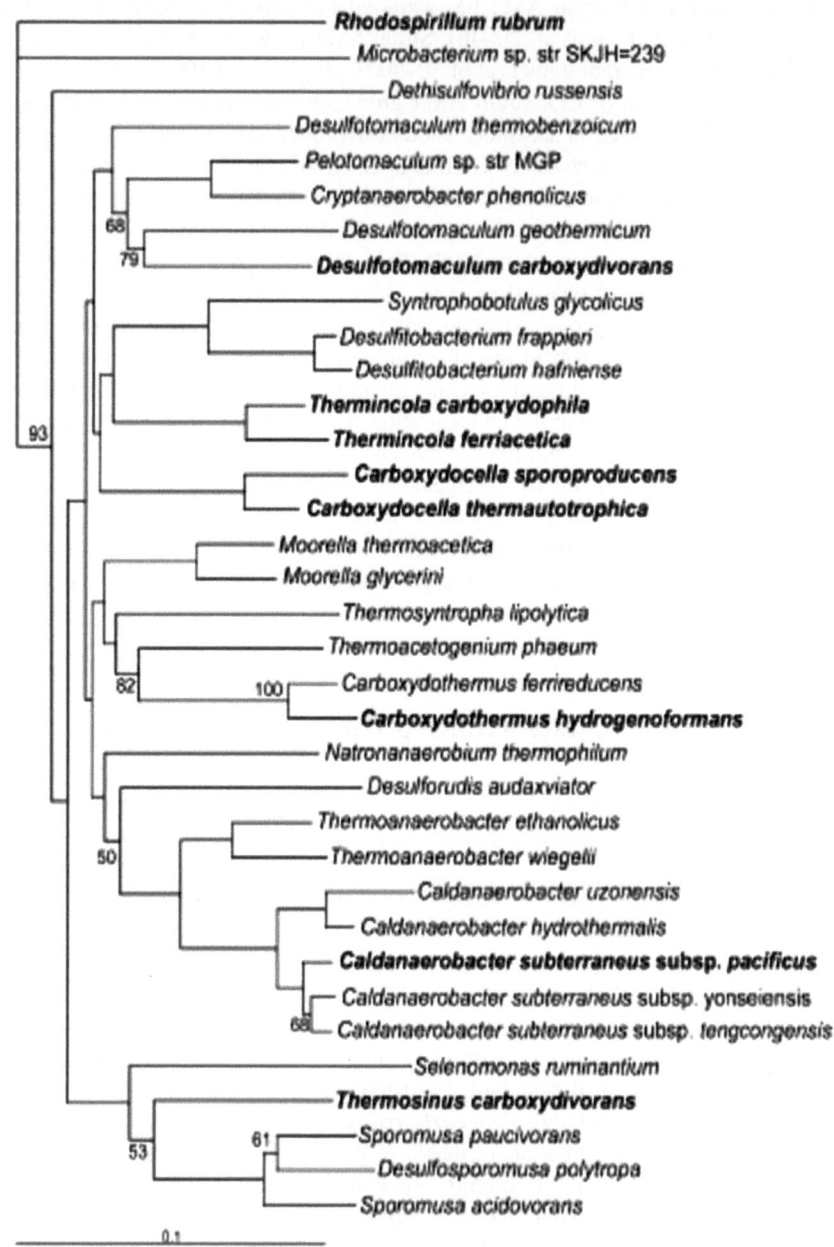

**Fig. 3.1** Phylogeny of hydrogenogens. A 16S rDNA tree was constructed in ARB using neighbor-joining methods. *Numbers* represent percentage of 1,000 bootstraps and are only shown for bootstraps greater than 50 %. *Species* in bold are known to perform hydrogenogenic carboxydotrophy (Techtmann et al. 2009)

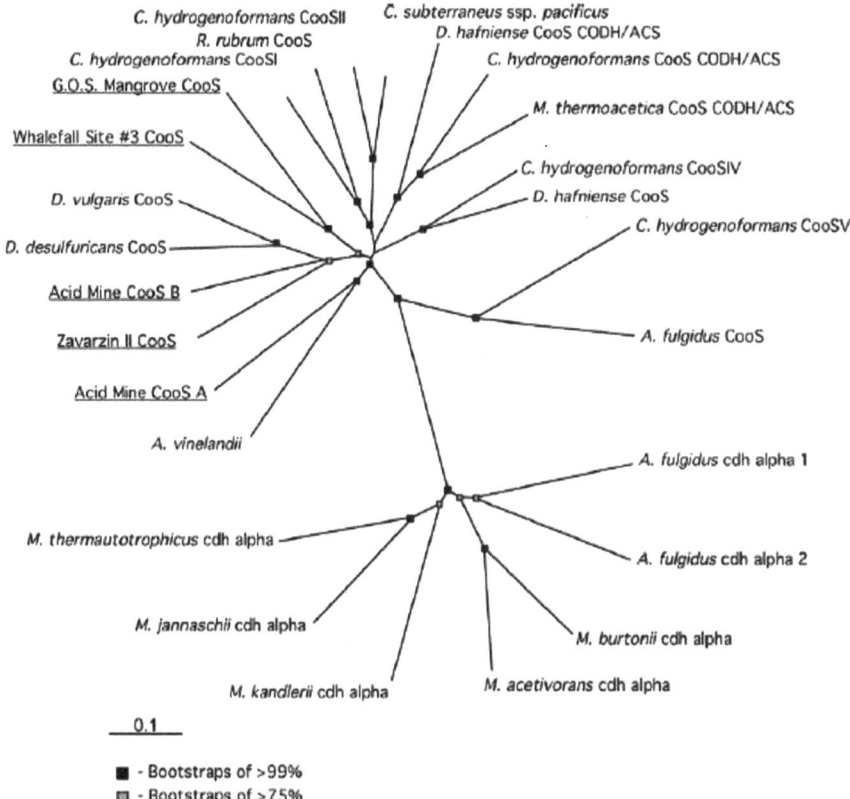

**Fig. 3.2** CODH catalytic subunit phylogeny: a phylogenetic tree comparing relatedness of the catalytic subunit from the CODH of various bacteria, archaea, and sequences from environmental DNA sequencing (Techtmann et al. 2009)

enzymatic steps as in bacterial sulfate reduction are employed (Rabus et al. 2006). The genome of *A. fulgidus* encodes several forms of CODH and one CODH/ACS (Klenk et al. 1997). It was thus not surprising that *A. fulgidus* is able to couple CO oxidation to sulfate reduction (Henstra et al. 2007a). However, it was very surprising that *A. fulgidus* is not strictly dependent on the presence of sulfate, thiosulfate, or other sulfur compounds as electron acceptor, but can grow chemolithoauthotrophically with CO as an acetogen (Henstra et al. 2007a). Thus, in the absence of sulfate, the only product of CO metabolism is acetate, which is formed via the reductive acetyl-CoA pathway with formyl-methanofuran as intermediate, and formate (Henstra et al. 2007a). *A. fulgidus* can also completely oxidize various organic compounds in the presence of sulfate or grow chemolithoautotrohically on $H_2$ with thiosulfate but not with sulfate as electron acceptor (Zellner et al. 1989).

## 3.3  CO-utilizing Sulfur-Reducing Archaea

*Thermoproteus tenax,* a hyperthermophilic crenarchaeota that can grow heterotro-phically by oxidizing sugars (Siebers and Schonheit 2005) or chemolithoautotro-phically by oxidizing $H_2$ and CO, coupled to the reduction of elemental sulfur producing hydrogen sulfide (Fischer et al. 1983). *T. tenax* is strictly dependent on the presence of sulfur as the terminal electron acceptor under lithotrophic conditions (Fischer et al. 1983).

## 3.4  CO-utilizing Acetogenic Bacteria

Acetogenic bacteria are obligate anaerobes that form acetate from $CO_2$ via reductive acetyl-CoA pathway (also known as Wood–Ljungdahl pathway) or from organic substrates (Wood 1991; Ljungdahl 1994; Drake et al. 2002). In the acetyl pathway, $CO_2$ molecules are reduced to methyl- and carbonyl-groups, and combined with CoA by CODH/ACS to form acetyl-CoA. Early studies on the physiology and biochemistry of acetogens showed that CO can be metabolized through this path-way according to the following reaction (Kerby and Zeikus 1983):

$$4CO + 2H_2O \rightarrow CH_3COO^- + 2CO_2 + H^+ \quad \Delta G° = -43.6\,kJ/mol\ CO$$

Some clostridia such as *Clostridium carboxidovorans* (an anaerobic mesophilic carboxydotroph) produce significant amount of ethanol, butyrate, and butanol besides acetate during growth on CO (Liou et al. 2005). Other acetate-producing carboxydotrophic clostria include *C. formiaceticum* (an anaerobic mesophilic car-boxydotroph) and *C. thermoaceticum* (an anaerobic thermophilic carboxydotroph) (Diekert and Thauer 1978). Among groups of anaerobic carboxydotrophs, the acetogens have attracted the most attention because they offer several promising routes to chemical and fuels production.

Many thermophilic acetogenic bacteria can also grow on CO. CO conversion for four thermophilic acetogens has been described by Balk et al. (2008). Among them are *M. thermoacetica, Moorella thermoautotrophica, Moorella perchloratiredu-cens,* and *Thermoanaerobacter kivui. M. thermoacetica,* and *M. thermoauto-trophica* can grow on CO at high partial pressures as sole energy source (Savage et al. 1987; Daniel et al. 1990). *M. thermoacetica* and *T. kivui* can oxidize CO during growth on other substrates (Sokolova and Lebedinsky 2013).

## 3.5   CO-utilizing Methanogenic Archaea

The methanogens are the best-studied *Archaea* able to grow with CO as the sole energy source (Daniels et al. 1977; O'Brien et al. 1984). Their principal energy metabolism, methanogenesis, proceeds via three distinct, but overlapping pathways: the $CO_2$-reducing pathway, the methylotrophic pathway, and the aceticlastic pathway (Thauer 1998; Deppenmeier et al. 1999). This process results in the formation of an ion motive force across the membrane (Fig. 3.3). With respect to substrate utilization, two major groups of methanogenic archaea can be distinguished: the hydrogeno-trophic methanogens and the methylotrophic methanogens (Boone et al. 1993). Hydrogenotrophic methanogens utilize to $H_2 + CO_2$ and formate as substrates. These organisms belong to the orders *Methanobacteriales*, *Methanococcales*, and *Met-hanomicrobiales*. Model organisms of this group are *Methanothermobacter ther-moautotrophicus* (formerly *Methanobacterium thermoautotrophicum* strain $\Delta$H), and *Methanococcus jannaschii*, of which the genomes have been sequenced com-pletely. The distantly related methylotrophic methanogens of the order *Methanos-arcinales* utilize simple C1 compounds such as methanol, methylamines, and methylthiols. Some of them are also able to grow on $H_2 + CO_2$ and on acetate (Boone et al. 1993). Most methanogens are known to possess the hydrogenotrophic pathway,

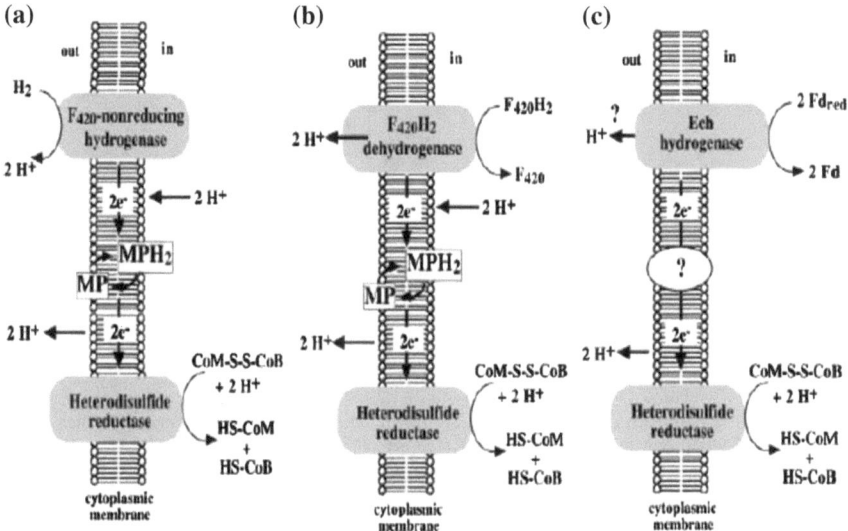

**Fig. 3.3** Membrane-bound electron transport in *Methanosarcina* strains. **a** $H_2$: heterodisulfide oxidoreductase system, **b** $F_{420}H_2$: heterodisulfide oxidoreductase system, **c** reduced ferredoxin: heterodisulfide oxidoreductase system; *MP* methanophenazine; *MPH$_2$*, reduced methanophen-azine; and *Fd$_{red}$* reduced ferredoxin (Deppenmeier 2002)

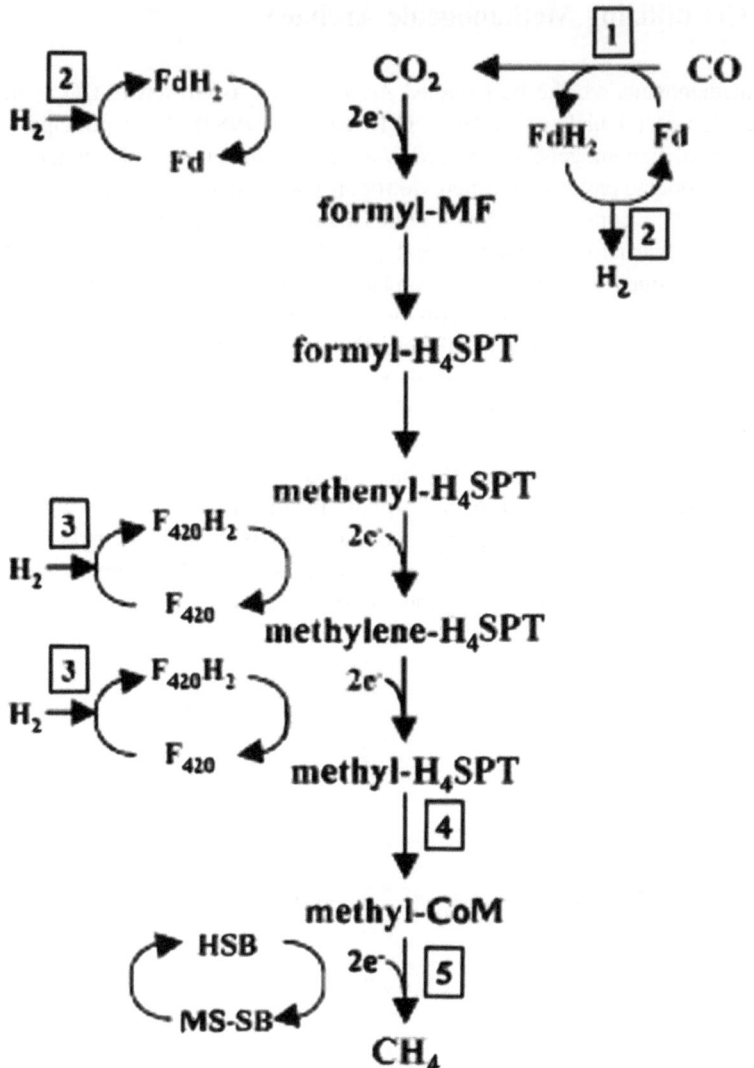

**Fig. 3.4** CO metabolism of methanogenic archaeon, *Methanosarcina barkeri*. Enzymes in this pathway include the following: (*1*), CODH(ACS); (*2*) Ech-hydrogenase; (*3*) $F_{420}$-dependent hydrogenase; (*4*) methyl-tetrahydrosarcinapterin:HSCoM methyltransferase; and (*5*) methyl-CoM reductase (Oelgeschlager and Rother 2008)

in which $CO_2$ is sequentially reduced to methane via coenzyme-bound intermediates using $H_2$ as the electron donor (Fig. 3.4).

The methanogens are the best-studied *Archaea* able to grow with CO as the sole energy source (Daniels et al. 1977; O'Brien et al. 1984). Their principal energy metabolism, methanogenesis, proceeds via three distinct, but overlapping pathways:

the $CO_2$-reducing pathway, the methylotrophic pathway, and the aceticlastic pathway (Thauer 1998; Deppenmeier et al. 1999). This process results in the formation of an ion motive force across the membrane (Fig. 3.3). With respect to substrate utilization, two major groups of methanogenic archaea can be distinguished: the hydrogenotrophic methanogens and the methylotrophic methanogens (Boone et al. 1993). Hydrogenotrophic methanogens use $CO_2$ as a source of carbon, and $H_2$ as a reducing agent. These organisms belong to the orders *Methanobacteriales*, *Methanococcales*, and *Methanomicrobiales*. Model organisms of this group are *M. thermoautotrophicus* (formerly *M. thermoautotrophicum* strain ΔH) and *M. jannaschii*, of which the genomes have been sequenced completely. The distantly related methylotrophic methanogens of the order *Methanosarcinales* utilize simple C1 compounds such as methanol, methylamines, and methylthiols. Some of them are also able to grow on $H_2 + CO_2$ and on acetate (Boone et al. 1993). Most methanogens are known to possess the hydrogenotrophic pathway, in which $CO_2$ is sequentially reduced to methane via coenzyme-bound intermediates using $H_2$ as the electron donor (Fig. 3.4).

CO has been a common methanogenic substrate. However, only *Methanothermobacter thermautotrophicus*, *Methanosarcina barkeri*, and *Methanosarcina acetovorans* were shown to utilize it for growth (Daniels et al. 1977; Krzycki et al. 1982; O'Brien et al. 1984; Rother and Metcalf 2004). In *M. barkeri* and *M. acetovorans,* four moles of CO are oxidized to $CO_2$ for every mole of $CO_2$ reduced to methane:

$$4CO + 2H_2O \rightarrow CH_4 + 3CO_2 \quad \Delta G° = -52.6\,kJ/mol\ CO$$

Both *M. thermautotrophicus* and *M. barkeri* grow slowly on CO (doubling time: >200 and 65 h, respectively) and growth ceases with increasing levels of this substrate. During carboxydotrophic growth, both organisms produce substantial amount of molecular hydrogen, which is eventually remetabolized. Therefore, carboxydotrophic growth of these two organisms can be considered hydrogenotrophic, combined with CO-dependent $H_2$ formation (Fig. 3.4). Deletion of the genes encoding Ech-hydrogenase (Fig. 3.3c), which is homologous to type III hydrogenase from E. coli (Hedderich and Forzi 2005), impairs growth of *M. barkeri* on $H_2 + CO_2$ and CO (Oelgeschlager and Rother 2008). *M. acetovorans*, which grows well on CO as the sole energy source, but not with $H_2 + CO_2$, does not encode Ech-hydrogenase (Fig. 3.2a). It is devoid of $H_2$ metabolism (Sowers et al. 1984) and CO-dependent $H_2$ formation has never been observed (Oelgeschlager and Rother 2008). Nevertheless, methane formation from CO appears to proceed via the common $CO_2$ reduction path (Fig. 3.5), since $CO_2$ is produced and because proteins known to be involved are also synthesized at elevated levels in CO-grown cells of *M. acetovorans* (Lessner et al. 2006; Rother et al. 2007).

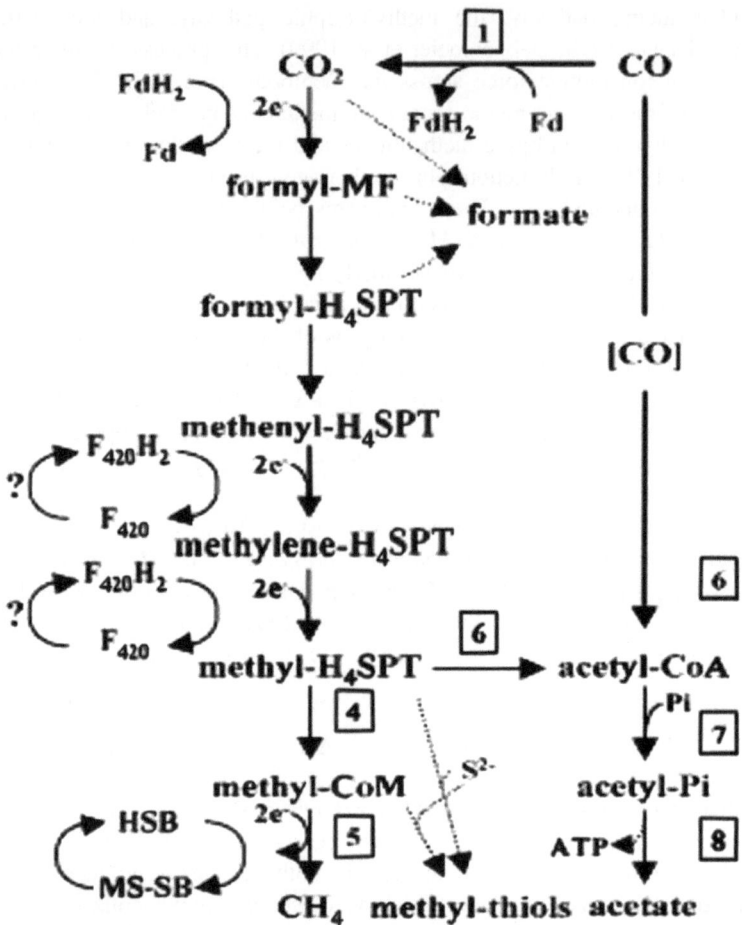

**Fig. 3.5** CO-metabolism of methanogenic archaeon, *Methanosarcina acetovorans*. Enzymes in this pathway include the following: (*1*), CODH(ACS); (*2*) Ech-hydrogenase; (*3*) $F_{420}$-dependent hydrogenase; (*4*) methyl-tetrahydrosarcinapterin:HSCoM methyltransferase; and (*5*) methyl-CoM reductase (*6*), CODH/ACS; (*7*) phosphotransacetylase; and (*8*), acetate kinase. *The dashed arrows* represent hypothetical reactions (Oelgeschlager and Rother 2008)

## 3.6  CO-utilizing Iron-Reducing Bacteria and Archaea

Dissimilatory iron reduction is found in many phylogenetic groups of bacteria or archaea (Weber et al. 2006). Examples of CO-utilizing iron-reducing microorganisms include *Carboxydothermus pertinax*, *Carboxydothermus siderophilus*, *Thermosinus carboxydivorans*, and *Thermincola ferriacetica* (Table 3.2).

C. pertinax (strain Ug1T), a novel anaerobic, Fe(III)-reducing, hydrogenogenic, carboxydotrophic bacterium, was isolated from a volcanic acidic hot spring in southern Kyushu Island, Japan (Yoneda et al. 2012). Cells of the isolate were rod-shaped (1.0–3.0 μm long) and motile due to peritrichous flagella. Strain Ug1T grew chemolithoautotrophically on CO (100 % in the gas phase) with reduction of ferric citrate, amorphous iron (III) oxide, 9,10-anthraquinone 2,6-disulfonate, thiosulfate, or elemental sulfur. No carboxydotrophic growth occurred with sulfate, sulfite, nitrate, or fumarate as electron acceptor. During growth on CO, $H_2$ and $CO_2$ were produced. Growth occurred on molecular hydrogen as an energy source and carbon dioxide as a sole carbon source. Growth was also observed on various organic compounds under an $N_2$ atmosphere with the reduction of ferric iron. The temperature range for carboxydotrophic growth was 50–70 °C, with an optimum at 65 °C. The pH (25 °C) range for growth was 4.6–8.6, with an optimum between 6.0 and 6.5. The doubling time under optimum conditions using CO with ferric citrate was 1.5 h. The DNA G+C content was 42.2 mol %. Analysis of 16S rRNA gene sequences demonstrated that this strain belongs to the thermophilic carboxydo-trophic bacterial genus *Carboxydothermus*, with sequence similarities of 94.1–96.6 % to members of this genus. The isolate can be distinguished from other members of the genus *Carboxydothermus* by its ability to grow with elemental sulfur or thiosulfate coupled to CO oxidation (Yoneda et al. 2012).

*C. siderophilus*, also known as strain 1315T, is an anaerobic, thermophilic, Fe (III)-reducing, CO-utilizing bacterium (Slepova et al. 2009). It was isolated from a hot spring of Geyser Valley on the Kamchatka Peninsula. Strain 1315T grew at 52–70 °C, with an optimum at 65 °C, and at pH 5.5–8.5, with an optimum at pH 6.5–7.2. In the presence of Fe(III) or 9,10-anthraquinone 2,6-disulfonate (AQDS), the bacterium was capable of growth with CO and yeast extract (0.2 g $L^{-1}$); during growth under these conditions, strain 1315T produced $H_2$ and $CO_2$ and Fe(II) or $AQDSH_2$, respectively. Strain 1315T also grew by oxidation of yeast extract, glucose, xylose, or lactate under an $N_{(2)}$ atmosphere, reducing Fe(III) or AQDS. Yeast extract (0.2 g $L^{-1}$) was required for growth. It grew exclusively with Fe(III) or AQDS as an electron acceptor. The generation time under optimal conditions with CO as growth substrate was 9.3 h. The G+C content of the DNA was 41.5 ± 0.5 mol %. 16S rRNA gene sequence analysis placed the organism in the genus *Carboxydothermus* (97.8 % similarity with the closest relative) (Slepova et al. 2009).

*T. carboxydivorans* (strain Nor1T) is an anaerobic, thermophilic, facultatively carboxydotrophic bacterium (Sokolova et al. 2004b). It was isolated from a hot spring at Norris Basin, Yellowstone National Park. Strain Nor1T was thermophilic (temperature range for growth was from 40 to 68 °C, with an optimum at 60 °C) and neutrophilic (pH range for growth was from 6.5 to 7.6, with an optimum at 6.8–7.0). It grew chemolithotrophically on CO, producing equimolar quantities of $H_2$ and $CO_2$. During growth on CO in the presence of ferric citrate or amorphous ferric iron oxide, strain Nor1T reduced ferric iron but produced $H_2$ and $CO_2$ at a ratio close to 1:1. Growth on CO in the presence of sodium selenite was accompanied by precipitation of elemental selenium. Elemental sulfur, thiosulfate, sulfate, and nitrate did not stimulate growth of strain Nor1T on CO, and none of these chemicals was reduced. Strain Nor1T was able to grow on glucose, sucrose, lactose, arabinose, maltose, fructose, xylose, and pyruvate, but not on cellobiose, galactose, peptone, yeast extract, lactate, acetate, formate, ethanol, methanol, or sodium citrate. Lactate, acetate, formate, and $H_2$ were not utilized either in the absence or in the presence of ferric iron, thiosulfate, sulfate, sulfite, elemental sulfur, or nitrate (Sokolova et al. 2004b).

*T. ferriacetica*, designated as strain Z-0001, is moderately thermophilic, spore-forming bacterium able to reduce amorphous Fe(III)-hydroxide (Zavarzina et al. 2007). It was isolated from ferric deposits of a terrestrial hydrothermal spring, Kunashir Island (Kurils). Strain Z-0001 was found to be an obligate anaerobe. It grew in the temperature range from 45 to 70 °C with an optimum at 57–60 °C; in a pH range from 5.9 to 8.0 with an optimum at 7.0–7.2; and in NaCl concentration range 0–3.5 % with an optimum at 0 %. Molecular hydrogen, acetate, peptone, yeast and beef extracts, glycogen, glycolate, pyruvate, betaine, choline, N-acetyl-D-glucosamine, and casamino acids were used as energy substrates for growth in the presence of Fe(III) as an electron acceptor. Sugars did not support growth. Magnetite, Mn(IV), and anthraquinone-2,6-disulfonate served as the alternative electron acceptors, supporting the growth of isolate Z-0001 with acetate as electron donor. Formation of magnetite was observed when amorphous Fe(III) hydroxide was used as electron acceptor. Isolate Z-0001 was able to grow chemolithoautotrophicaly with molecular hydrogen as the only energy substrate, Fe(III) as electron acceptor, and $CO_2$ as the carbon source. Isolate Z-0001 was able to grow with 100 % CO as the sole energy source, producing $H_2$ and $CO_2$, requiring the presence of 0.2 g $L^{-1}$ of acetate as the carbon source (Zavarzina et al. 2007).

# Chapter 4
# Biotechnological Applications of Thermophilic Carboxydotrophs

Microorganisms have been used in industry in a number of ways that generally exploit their natural metabolic capabilities. They are used in manufacture of foods and production of antibiotics, probiotics, drugs, vaccines, starter cultures, insecticides, enzymes, fuels, and solvents (Tiquia and Mormile 2010, 2013). In addition, with genetic engineering technology, bacteria can be programmed to make various substances used in food science, agriculture, and medicine. The possibilities of biotechnology are endless considering the gene reservoirs and genetic capabilities within the bacteria. Ever since the discovery of thermophilic microorganisms, scientists have been interested in the diversity of such extremophiles for various reasons including fundamental interests, biotechnological applications, and economic benefit (Charlier and Droogmans 2005; Spain and Krumholz 2011; Gugliandolo et al. 2012). Much attention has been paid on the molecular basis of thermoadaptation strategies involving cellular macromolecules. In recent years, thermophiles have been useful in degradation of complex organic molecules (Blumer-Schuette et al. 2008; Suryawanshi et al. 2010; Sizova et al. 2011), metal leaching (Chen and Pan 2010), production of compatible solutes (Empadinhas and da Costa 2006), water treatment technology (Liao et al. 2010), and production of biofuels (Wiegel 1980; Henstra et al. 2007b; Koskinen et al. 2008; Shaw et al. 2008; Taylor et al. 2009; Kongjan et al. 2010; Roberts et al. 2010).

Interest in applying carboxydotrophic microorganisms in suitable biocatalytic processes for the production of useful chemicals from CO gas has led to various studies into production of alternative compounds, e.g., methane, acetate, butyrate, and other organic compounds (Zeikus 1983). The conversion of CO to ethanol and butanol was demonstrated for *Butyribacterium methylotrophicum* (Shen et al. 1999) and for some *Clostridium* species strain P7 (Rajagopalan et al. 2002). Studies employing methyl viologen and other viologen dyes, as inhibitors of methanogenesis, resulted in the production of formate by *Methanosarcina barkeri* (Mazumder et al. 1985) and methanol by *Moorella thermoacetica* (White et al. 1987). Lapidus et al. (1989) showed that cell-free extracts of *Desulfovibrio*

© The Author(s) 2014                                                            29
S.M. Tiquia-Arashiro, *Thermophilic Carboxydotrophs and their Applications in Biotechnology*, Extremophilic Bacteria, DOI 10.1007/978-3-319-11873-4_4

*desulfuricans* can produce methanol, ethanol, acetic acid, and C8–C24 paraffins from CO to $H_2$ at elevated pressures. Incubations of different inocula with mixtures of $H_2/CO/CO_2$ revealed the presence of various acids, ranging from acetic to caproic acid, but only acetic acid and butyric acid were observed in high quantities (Levy et al. 1981). Table 4.1 summarizes thermodynamically possible synthesis gas fermentations, either from CO or $H_2/CO$. Besides the use of growing cells, also purified enzymes of CO converting organisms, especially CODH, might offer interesting potentials in biotechnology (Ferry 1995). Purified CODH could be used in biofilters for cleaning the air in underground car parks or in biosensors for CO detection (Colby et al. 1985). Recently, a CODH from *M. thermoacetica* was found to catalyze the reduction of 2,4,6-trinitrotoluene (TNT), an important chemical explosive commonly present in soil of military training sites (Huang et al. 2000). CODH can be applied in dechlorination as well as in the reductive carboxylation of phenols (Ferry 1995). With the increasing discovery of fast-growing anaerobic isolates capable of CO conversion together with the current interest in $H_2$, a biological alternative for the chemical water–gas shift reaction might represent one of the most interesting applications. In the next sections, the use of thermophilic carboxydotrophs in biotechnological applications such as in microbial fuel cell, synthesis bioenergy products, bioremediation, and detection of CO (biosensing tool) is described.

**Table 4.1** Summary of reported reactions with CO and $H_2/CO$

| Product | Reaction | $\Delta G^{o\prime}$ kJ mol $CO^{-1}$ |
|---|---|---|
| *Reactions from* CO | | |
| Acetate | $4CO + 2H_2O \rightarrow CH_3COO^- + H^+ + 2CO_2$ | −44 |
| Butyrate | $10CO + 4H_2O \rightarrow CH_3(CH_2)_2COO^- + H^+ + 6CO_2$ | −44 |
| Ethanol | $6CO + 3H_2O \rightarrow CH_3CH_2OH + 4CO_2$ | −37 |
| Formate | $CO + H_2O \rightarrow HCOO^- + H^+$ | −16 |
| Hydrogen | $CO + H_2O \rightarrow H_2 + CO_2$ | −20 |
| Methane | $4CO + 2H_2O \rightarrow CH_4 + 3CO_2$ | −53 |
| n-Butanol | $12CO + 5H_2O \rightarrow CH_3(CH_2)_3OH + 8CO_2$ | −40 |
| *Reactions from* $H_2/CO$ | | |
| Acetate | $2CO + 2H_2 \rightarrow CH_3COO^- + H^+$ | −67 |
| Butyrate | $4CO + 6H_2 \rightarrow CH_3(CH_2)2COO^- + H^+ + 2H_2O$ | −80 |
| Ethanol | $2CO + 4H_2 \rightarrow CH_3CH_2OH + H_2O$ | −72 |
| Methane | $CO + 3H_2 \rightarrow CH_4 + H_2O$ | −151 |
| Methanol | $CO + 2H_2 \rightarrow CH_3OH$ | −39 |
| n-Butanol | $4CO + 8H_2 \rightarrow CH_3(CH_2)_3OH + 3H_2O$ | −81 |

# 4.1 Electricity Production from CO/Syngas-fed Microbial Fuel Cell

## 4.1.1 Overview of Microbial Fuel Cells (MFCs)

Microbial fuel cells (MFCs) represent a novel technological solution for electricity production from biomass. In its most simple configuration, a microbial fuel cell is a device which uses microorganisms to produce an electrical current. The technology exploits the ability of microorganisms that are capable of extra-cellular electron transfer to an insoluble electron acceptor such as an electrode. Such microorganisms are commonly referred to as anodophilic, exoelectrogens, or electricigens, and the process is referred to as electrogenesis or electricigenesis (Logan 2008). The oxidation of organic chemicals by microorganisms liberates both electrons and protons. Electrons are then transferred from microorganisms to the anode and then subsequently to the cathode through an electrical network. Simultaneously, protons (electron acceptor) migrating to the cathode combine with electrons and an electron acceptor such as oxygen to produce water. The electrical current generated is similar to that in chemical fuel cells; however, in MFCs, microbial catalysts are attached to the anode surface (Franks and Nevin 2010).

MFCs are configured in a variety of configurations. Single-chamber MFCs are designed with an anodic compartment without the requirement for an aerated compartment containing cathode (Fig. 4.1). In a typical configuration, the anode contained in a compartment is coupled with a porous air-cathode (Shewa et al. 2014).

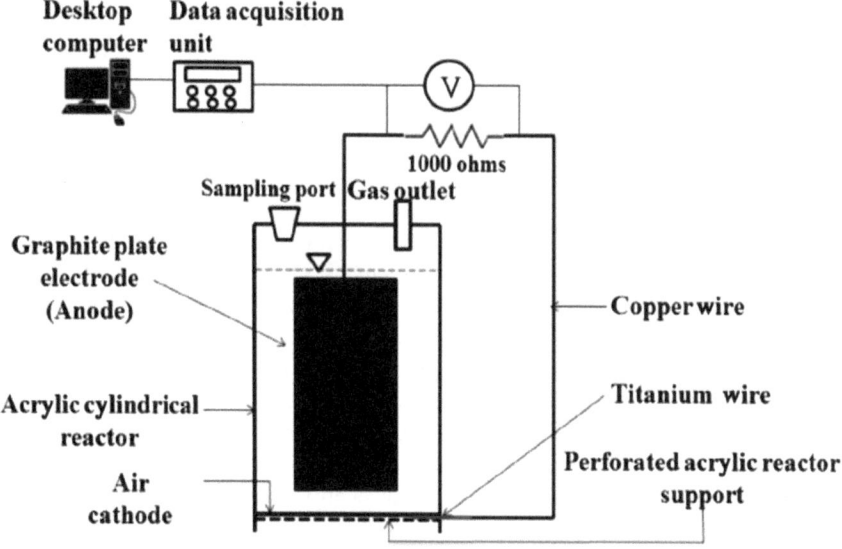

**Fig. 4.1** Schematic of a single-chamber microbial fuel cell (Shewa et al. 2014)

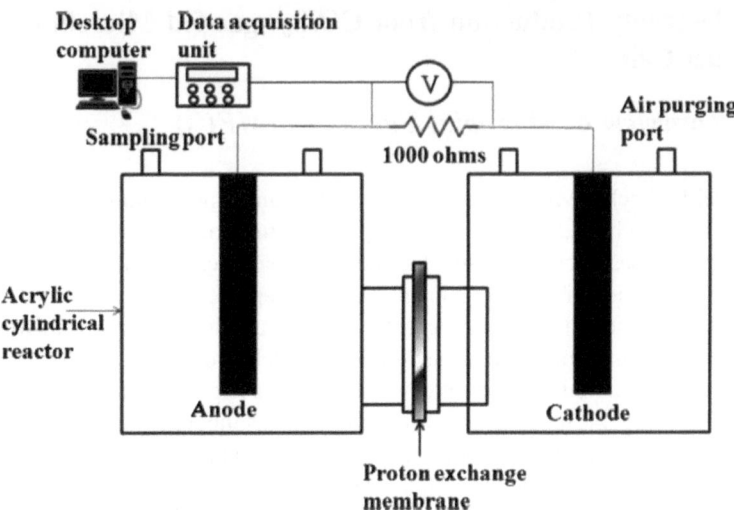

**Fig. 4.2**  Schematic of a two-chamber microbial fuel cell (Shewa et al. 2014)

In a two-chamber MFC configuration, the oxidation of the electron donors on an anode is physically separated from the reduction of an electron acceptor on the cathode. Microorganisms are cultivated on the anode where electron donors are oxidized. Electrons are transferred to the anode and subsequently to oxygen or other electron acceptors. Typically, the anode compartment is separated from the cathode compartment by a proton exchange membrane (PEM) or cation exchange membrane (CEM) (Fig. 4.2). The anodic chamber houses the electricigenic microorganisms which are laid upon a convoluted, conductive, and non-corrosive anode such as graphite brush, carbon felt, forming a biofilm. This biofilm first catalyzes the oxidation of the fed fuel (electron donor), liberating electrons and protons and then transferring the electrons to the anode (Rinaldi et al. 2008).

In the microbial fuel cell, substrate is oxidized at the anode part producing $CO_2$, proton ($H^+$) and electron. Electrons will go through external circuit to the cathode part producing current, and protons go through the exchange membrane. At the cathode part, oxygen reduction reaction occurs just like in the typical chemical fuel cell (Fig. 4.3). The role of the microorganisms in microbial fuel cell is very important. Some microorganisms cannot transfer the electron to the anode, and therefore, mediators such as thionine or methyl viologen have to be used to facilitate the electron transfer. However, recent development has shown that there exist electrochemically active bacteria such as *Shewanella putrefaciens*, *Aeromonas hydrophila*, which can transfer electron directly to the electrode without mediators.

In Fig. 4.3, it can be seen that the microorganisms in microbial fuel cell convert the biomass substrate directly into electrical energy, while in the biogas fermentation method (Fig. 4.4), biomass substrate is degraded into eventually the mixture of $CH_4$ and $H_2$ (chemical fuels). Therefore, appropriate kinds of microorganisms

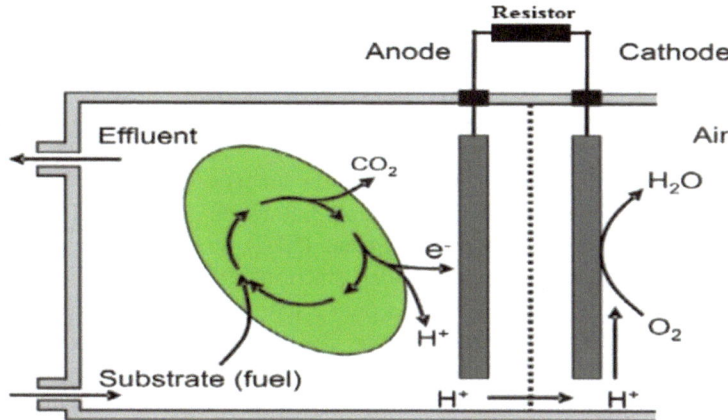

**Fig. 4.3** A fermentative microorganism converts substrate to end products, $CO_2$, and hydrogen. The hydrogen can abiotically react with the anode to produce electrons and protons. This process only partially recovers the electrons available in the organic fuel as electricity and results in the accumulation of organic products in the anode chamber (Logan 2008)

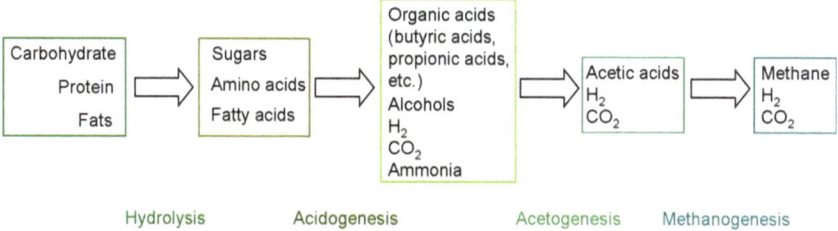

**Fig. 4.4** Biological and chemical stages of biogas fermentation processes. The exact chemical species produced at each stage depend considerably upon the kinds of microorganisms used as well as processing conditions

used in microbial fuel cell are very crucial since they determine the actual mechanism of the oxidation reactions as well as mechanism of electron transfer in the anode part of the fuel cell, which will influence how effectively we can obtain electrical energy from this system. As an example, the oxidation of glucose in the anodic chamber is presented in equation below:

$$C_6H_{12}O_6 + 6H_2O \rightarrow 6CO_2 + 24H^+ + 24e^-$$

While the electrons travel through an electrical load (device to be powered or a resistor in laboratory-scale studies) and generate electricity until reaching the cathode, the corresponding protons migrate through the separator to the cathodic compartment to maintain the electrical neutrality. At the cathode, the protons combine with the electrons and an electron acceptor (catholyte) such as oxygen, to form $H_2O$ through a reduction reaction as shown in equation below:

$$6O_2 + 24H^+ + 24e^- \rightarrow 12H_2O$$

Generating power in MFCs depends on oxidation–reduction (redox) chemistry. MFCs contain anodic and cathodic compartments, each of which holds an electrode separated by a cation-permeable membrane (Fig. 4.5). In the anode chamber, microbial substrates such as acetate (an electron donor) are oxidized in the absence of oxygen by respiratory bacteria, producing protons, and electrons. The electrons are passed through an electron transport chain (ETC), and protons are translocated across the cell membrane to generate adenosine triphosphate (ATP). Electrons and protons exiting the ETC typically pass onto a terminal electron acceptor such as oxygen, nitrate, or Fe(III). However, in the absence of such acceptors in an MFC, some microorganisms pass the electrons onto the anode surface. Difference in redox potentials (i.e., the ability of a compound to donate or accept electrons, denoted $E_o$ and measured in volts) between the electron donor and the electron acceptor is the determinant of the potential energy available to the microorganism for anabolic processes. In an MFC, the electrochemical redox potential difference of the anode and cathode determines how much energy is available. Electrons produced in an MFC flow from the anode through an external electrical circuit to the cathode to generate electrical current. While electrons move externally, protons diffuse from the anode to the cathode via the cation membrane to complete the internal circuit (Fig. 4.5). At the cathode, the electrons and protons combine to reduce the terminal electron acceptor, which in many applications is oxygen. Therefore, bacteria in the anode are physically separated from their terminal electron acceptor in the cathode compartment. The electrical power (measured in watts) produced by an MFC is based on the rate of electrons moving through the circuit (current, measured in amps) and electrochemical potential difference (volts) across the electrodes. Many factors affect current production, including substrate concentration, bacterial substrate oxidation rate, presence of alternative electron acceptors, and microbial growth. Electrochemical potential, on the other hand, depends on the redox couple between the bacterial respiratory enzyme or electron carrier and the potential at the anode, which is determined by the terminal electron acceptor in the cathode and any system losses (Wrighton and Coates 2009).

For bacteria to produce electricity in MFCs, the cells need to transfer electrons generated along their membranes to their surfaces. Yet very little is known about bacterial interactions with electrodes. While anodes and cathodes can function in bacterial respiration, research has been focused on understanding microbial anodic electron transfer. Anode-respiring bacteria catalyze electron transfer in organic substrates onto the anode as a surrogate for natural extracellular electron acceptors (e.g., ferric oxides or humic substances) by a variety of mechanisms (Fig. 4.6). Bacteria transfer electrons to anodes either directly or via mediated mechanisms. In direct electron transfer, bacteria require physical contact with the electrode for current production. The contact point between the bacteria and the anode surface requires outer membrane-bound cytochromes or putatively conductive pili called nanowires. Although direct contact of an outer membrane cytochrome to an anodic

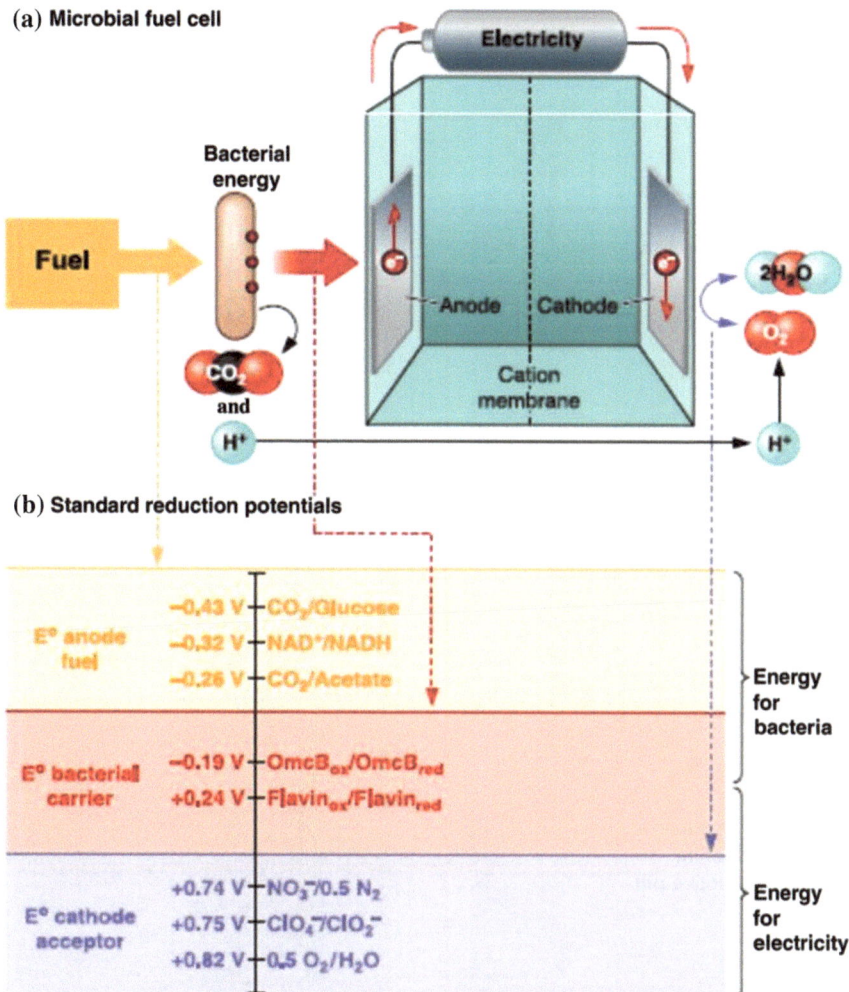

**Fig. 4.5 a** Schematic of a microbial fuel cell illustrating oxidation of fuel by bacteria in the anode compartment to produce electrons and protons. Electrons (*red arrow*) resulting from microbial oxidation flow from the anode through the external connection to the cathode to generate electrical current, while protons (*green arrow*) travel through the cation membrane. Together, both function to reduce the terminal electron acceptor, which in this case is oxygen, in the cathode. **b** Redox tower for components that are significant to current production in a MFC. For a redox couple, the electron donor must have a greater negative potential than the electron acceptor; this difference in potential is proportional to the amount of energy generated from the reaction. Current generation in a MFC is based on sequential redox reactions. First, bacteria oxidize fuel (potentials noted in *blue*) and transfer these electrons to an electron carrier at a more positive potential (noted in *red*), thereby generating energy for the bacteria. The final power generated by an MFC is based on the current production and redox couple between the bacterial respiratory enzyme or electron shuttle and the potential at the anode, which is determined by the terminal electron acceptor in the cathode and any system losses (Marsili et al. 2008)

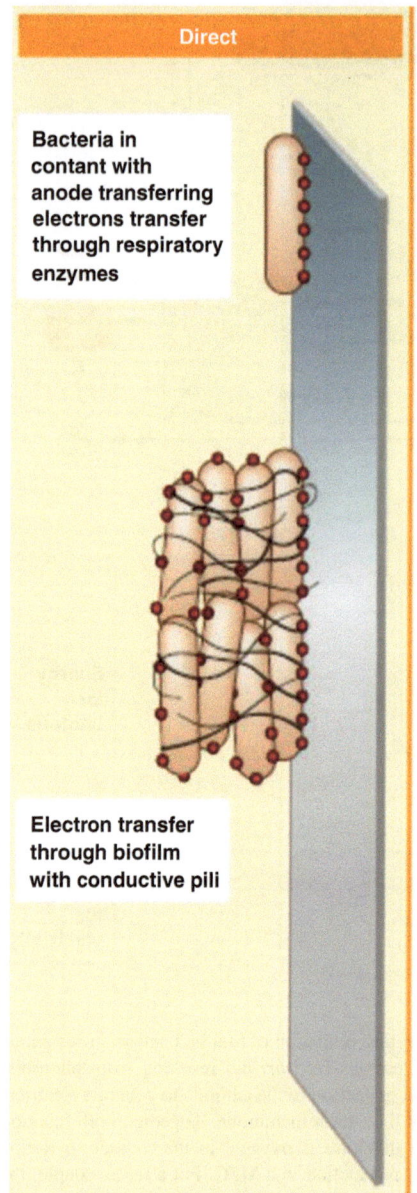
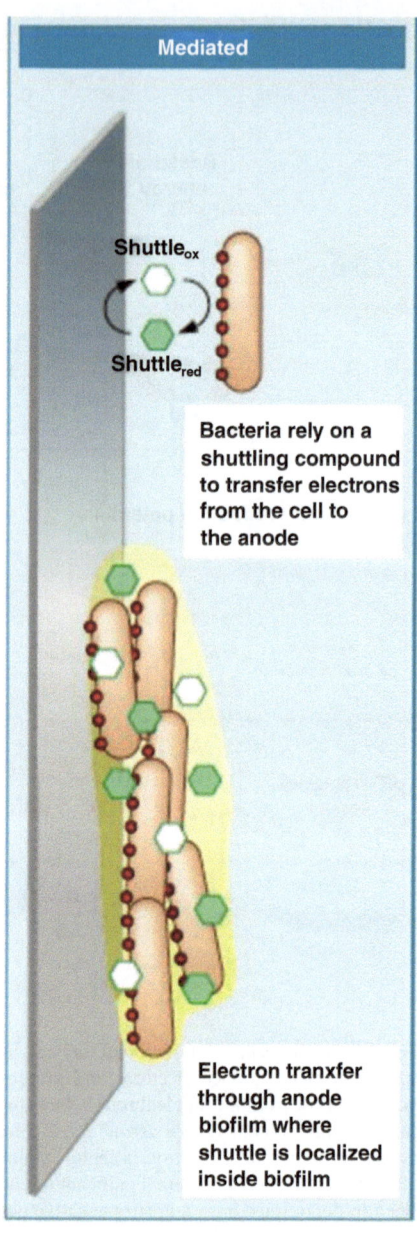

**Direct**

Bacteria in contant with anode transferring electrons transfer through respiratory enzymes

Electron transfer through biofilm with conductive pili

**Mediated**

Shuttle$_{ox}$

Shuttle$_{red}$

Bacteria rely on a shuttling compound to transfer electrons from the cell to the anode

Electron tranxfer through anode biofilm where shuttle is localized inside biofilm

**Fig. 4.6** Bacteria use direct and mediated mechanisms to transfer electrons from the cell membrane to the anode surface. Electrons can be transferred from the cell or through a conductive biofilm using each method (Lovley 2008)

surface would require microorganisms to be situated upon the electrode itself, direct electron transfer mechanisms are not limited to short-range interactions, as nano-wires produced by *Geobacter sulfurreducens* have been implicated in electron conduction through anode biofilms more than 50 mm thick. In mediated electron transfer mechanisms, bacteria either produce or take advantage of indigenous soluble redox compounds such as quinones and flavins to shuttle electrons between the terminal respiratory enzyme and the anode surface.

MFCs have been successfully operated on a wide range of substrates such as acetate, CO, $H_2$, glucose, galactose, butyrate, starch, marine sediments, and swine wastewater. (Pant et al. 2010; Hussein et al. 2011). In principle, any biodegradable material could be utilized as a fuel for electricity generation in an MFC (Fig. 4.7). An ideal MFC can produce current while sustaining a steady voltage as long as a steady supply of substrate in maintained. The theoretical ideal voltage, $E_{ideal}$ (V), attainable in an MFC can be thermodynamically predicted by the Nernst equation:

$$E_{ideal} = E_0 - (RT/nF) \ln(\Pi)$$

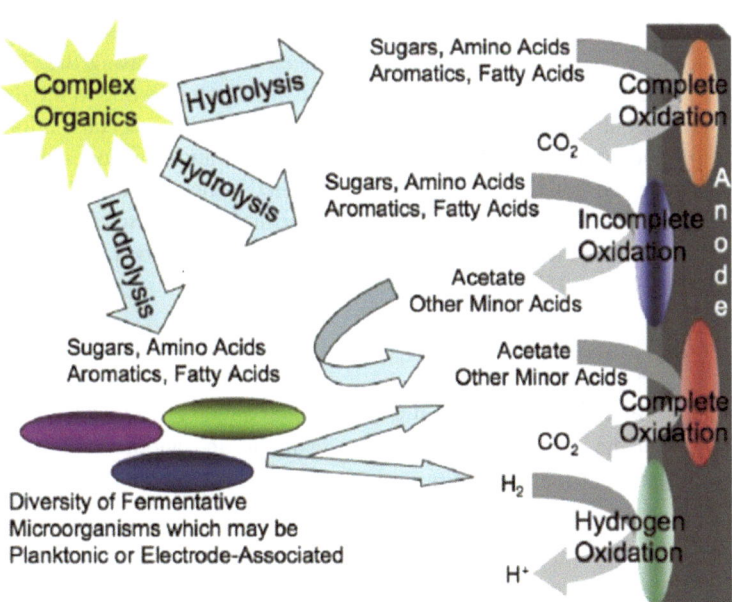

**Fig. 4.7** Simplified model for the conversion of complex organic fuels to electricity. Complex organic matter is hydrolyzed to constituents, in which most cases are primarily fermented, but there are microorganisms that can completely oxidize such compounds with an electrode serving as the sole electron acceptor or incompletely oxidize these substrates with electron transfer to an electrode. Acetate and other more minor fermentation acids can be completely oxidized to carbon dioxide, and this will typically be the primary source of electrons for current production. Hydrogen produced from fermentation may also be a source of electrons. Direct electron transfer to the anode is illustrated, but indirect electron transfer to the anode via soluble electron shuttles is also possible (Lovley 2008)

where $E_0$ is the standard cell potential (V), $R$ is the universal gas constant ($8.314$ J mol$^{-1}$ K$^{-1}$), $T$ is the temperature (K), $n$ is the number of electrons transferred in the reaction (dimensionless), $F$ is the Faraday's constant ($96{,}485$ C mol$^{-1}$), and $\Pi$ is the chemical activity of the products divided by those of the reactants (dimensionless). In practice, the actual voltage attainable in an MFC is less than the predicted voltage due to various irreversible losses or overpotentials. Activation, ohmic, and mass transport losses are the three major irreversible losses that affect MFC performance (Rismani-Yazdi et al. 2008). Briefly, activation losses are due to the activation energy that must be overcome by the reacting species at each electrode and largely depends on the electrochemical properties of the deployed electrodes, current density of the anode, operating temperature, and presence of electrochemical mediators. Hence, such voltage losses could be minimized by improving the electrode configuration and surface area, increasing the operating temperature and by utilizing microorganisms with high bioelectrochemical activity or enriching the electrode chambers with electromediating compounds (Rinaldi et al. 2008). Ohmic losses can be broadly ascribed to the electronic flow through the electrodes, the current collectors and the contact, and also to the resistance to the flow of ions through the electrolyte. Such losses could be reduced by improving the reactor design to reduce the distance between the electrodes, utilization of PEMs (if used) with low resistance and by increasing the electrolyte conductivity. Concentration losses represent the voltage losses due to the depletion of the reactants in the electrolyte near the electrodes and the accumulation of the reaction products. Improvement in the reactor configuration and operating parameters to reduce the concentration gradient minimizes such losses. The actual cell voltage, $E$cell (V), of an MFC can thus be determined by subtracting the voltage losses in anodic and cathodic compartment and can be described by the following equation (Rinaldi et al. 2008):

$$E\text{cell} = \left[ E_{\text{cathode}^-} \left| \eta_{\text{act},c} + \eta_{\text{conc},c} \right| \right] - \left[ E_{\text{anode}^-} \left| \eta_{\text{act},a} + \eta_{\text{conc},a} \right| \right] - \eta_{\text{ohm}}$$

where $E_{\text{cathode}}$ and $E_{\text{anode}}$ represent the cathode and anode potentials (V), $\eta_{\text{act},c}$ and $\eta_{\text{act},a}$ represent the activation losses in the cathodic and anodic chamber, $\eta_{\text{conc},c}$ and $\eta_{\text{conc},a}$ represent the concentration losses in the cathodic and anodic chamber, and $\eta_{\text{ohm}}$ represents the ohmic losses.

### 4.1.2 CO/Syngas as Substrates for Microbial Fuel Cells (MFCs)

The gasification of biomass at high temperatures leads to the generation of synthesis gas (syngas). CO and $H_2$ account for 60 % of the syngas composition, with $CH_4$, $CO_2$, $SO_2$, $H_2S$, and $NH_3$ present in smaller amounts (Sipma et al. 2006; Munasinghe and Khanal 2010). Although most of the syngas today are produced

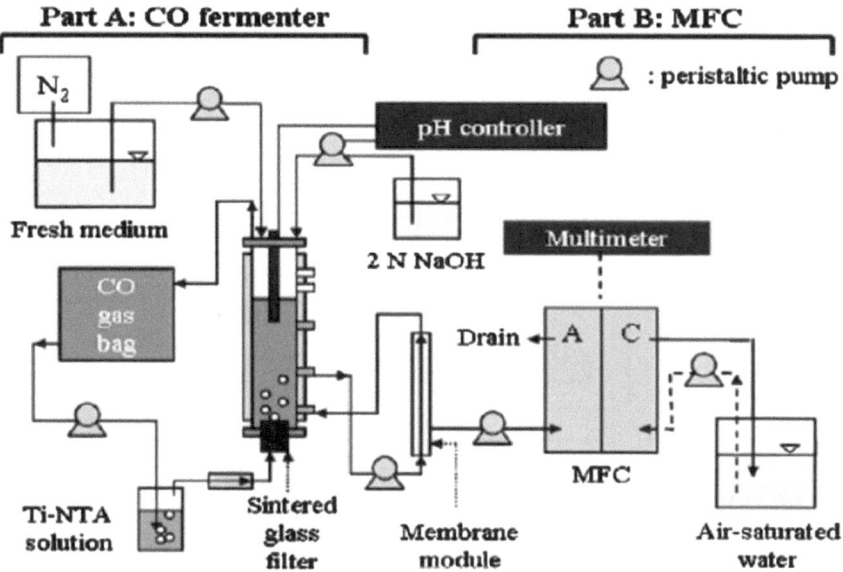

**Fig. 4.8** Schematic of the two-stage process for electricity production from carbon monoxide utilized in the study of Kim and Chang (2009)

from non-renewable resources, such as natural gas and coal, syngas production from renewable biomass or poorly degradable organic matter makes energy generation from syngas a sustainable process (Faaij et al. 1997; Henstra et al. 2007b). Fermentation is one way of producing an energy carrier from syngas, which has been advocated by many studies (Bredwell et al. 1999; Sipma et al. 2006; Henstra et al. 2007b). Also, electricity generation from syngas has been studied (Steele et al. 1999; Baschuk and Li 2001; Song 2002; Ormerod 2003; Hussain et al. 2011). The possibility of electricity production from CO and syngas in an MFC has been demonstrated by Kim and Chang (2009) in a two-stage reactor system as shown in Fig. 4.8, in which CO was first microbiologically converted to fermentation products (dominantly acetate), which were subsequently fed to an MFC seeded with an anaerobic sludge. A maximum power output of 1.6 mW $L^{-1}$ (normalized to the total reactor volume) and Coulombic efficiency (CE) of $\sim 5$ % were reported.

Mehta et al. (2010) demonstrated electricity generation in an MFC directly fed with CO/syngas. The maximum volumetric power output and CE of the system were 6.4 mW $L^{-1}$ and 8.7 %, respectively. Though the overall performance of the MFC directly fed with CO/syngas was only marginally better than the two-stage process utilized in the study of Kim and Chang (2009), it clearly demonstrated that the microbial communities of an MFC could utilize CO/syngas as an electron donor for electricity generation. The low performance of the MFC on CO/syngas was attributed to the inefficient system design and gas delivery. The utilization of a sparger system for gas delivery (Fig. 4.9) not only required the doubling of the reactor volume which invariably resulted in low power density, but also the

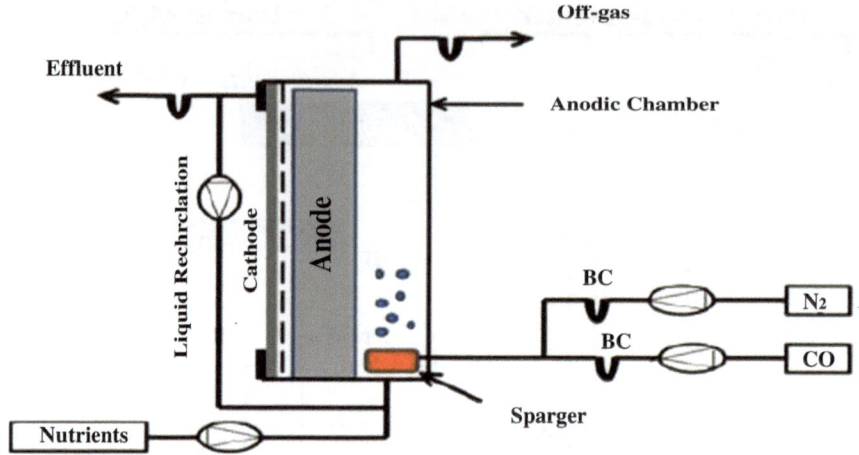

**Fig. 4.9** Schematic of the experimental setup for electricity generation from carbon monoxide/syngas utilized by Mehta et al. (2010)

inefficient gas–liquid mass transfer required much higher CO/syngas feed in rates to reach the desired dissolved CO levels; resulting in considerable gas losses and low CO transformation efficiency. This suggested that the performance of an MFC on CO/syngas could be enhanced by adoption of an efficient gas–liquid mass transfer mechanism and reactor design optimization.

Based on the analysis of metabolic products, Mehta et al. (2010) concluded that the production of electricity from CO or syngas in an MFC proceeds through a multi-step biotransformation process. Several concurrent pathways, as shown in Fig. 4.10, were hypothesized: One pathway involved CO transformation to acetate by acetogenic carboxydotrophic (CO-oxidizing) microorganisms followed by oxidation of acetate by CO-tolerant electricigenic microorganisms (pathway 1–4 in Fig. 4.10). This pathway was hypothesized to be the foremost responsible for

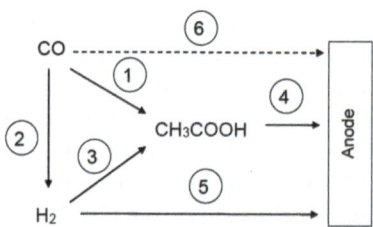

**Fig. 4.10** Proposed pathways of electricity production from CO and syngas in an MFC (Mehta et al. 2010). *Notations* (*1*) CO conversion to acetate by acetogenic carboxydotrophs; (*2*) CO conversion to $H_2$ by hydrogenogenic carboxydotrophs; (*3*) $H_2$ conversion to acetate by homoacetogens; (*4*, *5*) acetate and $H_2$ consumption by electricigenic microorganisms; and (*6*) CO consumption by electricigenic carboxydotrophs (hypothesized)

electricity generation. Acetate injection in the MFC always led to a dramatic increase in power production. In addition to acetate, other degradation products such as hydrogen were also found in the off-gas samples during MFC operation solely on CO, which indicated the presence of hydrogenogenic carboxydotrophic microorganisms. It was hypothesized that $H_2$ is also utilized for electricity production by the electricigenic microorganisms (pathway 3–5 in Fig. 4.10). The importance of this pathway might be increased in the syngas-fed MFC where a significant amount of hydrogen is present. Notably, the ability of the electricigenic microorganisms to utilize $H_2$ as an electron donor has been documented (Bond and Lovley 2003). The presence of acetate when the MFC was fed only with hydrogen also indicated the presence of homoacetogenic microorganisms. Based on this observation, a pathway of electricity production through $H_2$ and acetate followed by acetate conversion to electricity was also suggested (pathway 2–3–4 in Fig. 4.10). The work of Mehta et al. (2010) also raised a possibility of another pathway, which involved direct electron transfer to the anode by the metal-reducing carboxydotrophic bacteria (pathway 5 in Fig. 4.10). Microorganisms capable of utilizing CO as an electron donor for metal reduction such as *Thermosinus carboxydivorans* and *Carboxydothermus ferrireducens* have been isolated recently (Sokolova et al. 2004).

Based on the experimental observations from the studies of Kim and Chang (2009) and Mehta et al. (2010), it is evident that the electricity production from syngas in an MFC poses a number of engineering and microbiological challenges pertaining to gas transfer limitations, selection, and enrichment of microorganisms capable of efficient syngas transformation to electricity, and selection of cathodic catalysts resistant to poisoning by CO and sulfur compounds. The subsequent sections of this chapter review the microbial communities and reactor designs suitable for MFC operation on CO/syngas.

### 4.1.3 Thermophilic Carboxydotrophs in CO/Syngas-fed Microbial Fuel Cells (MFCs)

The search for microorganisms that are capable of catalyzing the reduction of an electrode within a fuel cell has primarily been focused on bacteria that operate at mesophilic temperatures. However, anaerobic digestion studies have reported on the superiority of thermophilic operation and demonstrated a net energy gain in terms of methane yield. Bacteria that function optimally under extreme conditions are beginning to be examined because they may serve as more effective catalysts (higher activity, greater stability, longer life, and capable of utilizing a broader range of fuels) in MFCs. CO/syngas-fed MFCs have been so far operated at mesophilic temperatures. Considering that at the exit of the gasification process syngas temperature could be in a range of 45–55 °C, the operation of the MFC at thermophilic temperatures might be preferable because it eliminates the need for syngas cooling and might lead to a higher biocatalytic activity (Jong et al. 2006; Mathis et al. 2008). The thermophilic conditions would also lead to reduced oxygen solubility, which is beneficial considering that even trace amounts of unreacted $O_2$ diffusing through the cathode can inhibit the anaerobic carboxydotrophic microorganisms which are highly sensitive to the presence of $O_2$ (Daviddova et al. 1994). This section discusses thermophilic microbial communities that could be utilized in an MFC for efficient CO/syngas transformation to electricity based on the biochemical pathways discussed in the previous section and as illustrated in Fig. 4.10.

Hussain et al. (2012) demonstrated for the first time that electricity can be generated in a thermophilic MFC fed with syngas. The thermophilic conditions led to a higher power density and improved syngas transformation efficiency as compared to a similar MFC operated under mesophilic conditions. The CE was also improved to 20–26 % as compared to 6–9 % reported for the mesophilic MFC. The supply of CO to the anodic liquid was also improved (Hussain et al. 2012). Several reasons can be cited to explain the improved MFC performance under thermophilic conditions. Firstly, thermophilic conditions affect the activation, ohmic, and diffusion losses at the anode. The activation losses contribute to 5–10 % of the total internal resistance in a mesophilic MFC (Logan 2008; Zhang and Liu 2010). Secondly, the electrochemical reaction rates increase with increasing temperature, thus leading to lower activation losses. Thirdly, temperature affects the diffusion of the substrates in the anodic liquid, thereby influencing the concentration losses, which account for 45–50 % of the total internal resistance of a MFC (Logan 2008; Zhang and Liu 2010). Likely, the operation of the syngas-fed MFC at thermophilic temperatures increased the transfer rate of CO and $H_2$ not only to the anodic liquid, but also facilitated the transport of the dissolved gasses through the stagnant liquid layer adjacent to the anode fibers, thus reducing the diffusion losses. Overall, the internal resistance of the thermophilic MFC at optimized performance was less than 246.50 $\Omega$, whereas the mesophilic MFC had an internal resistance of above 120 $\Omega$ (Mehta et al. 2010; Hussain et al. 2011). Finally, thermophilic conditions increased the activity of the carboxydotrophic microorganisms. Up to a certain temperature, the biomass growth and substrate conversion rates increase with temperature according to the Arrhenius relationship. In general, the thermophilic microorganisms feature higher growth and reaction rates as compared to the mesophilic cultures (Min et al. 2008). Therefore, a higher carboxydotrophic activity could be expected at thermophilic temperatures. In a mesophilic CO-fed MFC, the step of CO conversion to acetate appeared to limit the overall transformation rate (Mehta et al. 2010). Another advantage could be related to the reduced $O_2$ solubility at elevated temperatures. While most of the $O_2$ diffusing through the cathode surface is consumed by the cathodic reaction, the residual $O_2$ diffuses to the anodic liquid thus resulting in the inhibition of the anodophilic and carboxydotrophic populations (Oelgeschlager and Rother 2008) as well as competing with the anode as the final electron acceptor (Liu et al. 2004). The presence of trace amounts of $O_2$ in the anodic chamber was observed to significantly impair the power output of the mesophilic syngas-fed MFC (Hussain et al. 2011).

### 4.1.3.1 Syngas Conversion to Electricity by Thermophilic Acetogenesic and Electricigeneic Carboxydotrophs

Many acetogens capable of growth on CO under thermophilic conditions have been reported in Sect. 3.4. Savage et al. (1987) reported the ability of CO-dependent chemolithotrophic acetogenesis and growth by *Moorella thermoautotrophicum* (earlier *Clostridium thermoautotrophicum*), with supplemental $CO_2$ required for efficient growth on CO. The $CO/CO_2$ ratio of 2:4 yielded optimal doubling times at a temperature of 58 °C. This microorganism has the ability to grow autotrophically and heterotrophically using various electron donors and acceptors (Savage et al. 1987; Sokolova et al. 2009). *Clostridium thermoaceticum* also demonstrated CO-dependent growth and acetogenesis under chemolithotrophic conditions with a doubling time of 10 h at a temperature of 55 °C (Daniel et al. 1990). The recently isolated thermophilic bacterium *Moorella perchloratireducens*, which is closely related to the above-mentioned bacterial species, resorted to acetogenesis in the absence of perchlorate (Balk et al. 2008). Acetogenic hydrogenation or the conversion of $H_2$ and $CO_2$ to acetate has also been observed in thermophiles. *C. thermoaceticum*, capable of growth on CO, grew chemolithotrophically on $H_2$ and $CO_2$ forming acetate. The doubling time was 18 h at a temperature of 55 °C (Kerby and Zeikus 1983). A similar growth physiology was also observed in *Acetogenium kivui* (Daniel et al. 1990). This thermophilic anaerobic bacterium formed acetate by chemolithotrophic growth on $H_2$ and $CO_2$ with a doubling time of 2 h, at a temperature of 66 °C and pH of 6.8. For generation of electricity in an MFC, these thermophilic carboxydotrophic microorganisms have to be co-cultured with thermophilic electricigenic microorganisms capable of using acetate as an electron donor. Although the microorganisms studied for generation of electricity in an MFC are predominantly mesophilic (e.g., *G. sulfurreducens* or *Geobacter metallireducens*), successful MFC operation under thermophilic conditions has been demonstrated. Choi (2004) used thermophilic bacteria *Bacillus licheniformis* and *Bacillus thermoglucosidasius* (with a redox mediator) for electricity generation. The best efficiency was achieved within the temperature range of 50–60 °C. MFC operation under thermophilic conditions was also reported by Jong et al. (2006). In their study, the MFC was inoculated with anaerobic digester effluent and fed with sodium acetate. The maximum power density was achieved during MFC operation at 55 °C. Based on the 16S rRNA analysis, only 13 different patterns of anodic bacterial populations were observed, of which 5 patterns showed the highest homology to an uncultured clone E4, which was initially identified as a member of a thermophilic microbial community in a laboratory-scale methanol-fed anaerobic digester. Seven patterns were related to genus *Coprothermobacter* and one pattern was related to *Thermodesulfovibrio* spp.

Mathis et al. (2008) studies thermophilic bacteria selected from sediment MFC were used to colonize the anode of acetate- and cellulose-fed MFCs. Cloning and sequencing of the biofilm formed at the anode of the acetate-fed MFC showed the presence of *Deferribacters* and *Fermicutes*. Interestingly, 48 clones (out of 64) of *Fermicutes* had RFLP patterns and sequences (99 %) most similar to that of *Thermincola carboxydophila*, a hydrogenogenic CO-oxidizing thermophilic microorganism (Mathis et al. 2008). *Firmicutes* spp. are also identified during thermophilic MFC operation by Wringhton et al. (2008). This study provided a detailed analysis of microbial community dynamics in an acetate-fed MFC inoculated with sludge collected from a thermophilic anaerobic digester. The dominant members of the electricity producing community were identified using clone library analysis. The results showed the dominance of *Firmicutes* spp. (80 % of the clone library sequences). Within *Firmicutes*, sequences belonging to *Thermicanus, Alicyclobacillus, and Thermincola* were identified representing 27, 25, and 22 % of the total clones, respectively. The study was well complemented by the demonstration of electricity production in an MFC inoculated with a pure strain of *Thermincola sp.* Strain JR, representing direct anode reduction by a member of *Fermicutes* phylum (Wrighton et al. 2008).

Since electricity generation has never been an evolutionary pressure per se, but rather the capacity for electron transfer to natural extracellular electron acceptors, it is likely that the ability of the microorganism to produce electricity is closely correlated to their capacity to transfer electrons onto extracellular acceptors, such as Fe(III) and Mn(IV) oxides, and humic substances (Lovley et al. 2004; Lovley 2006; Logan 2009). Hence, the prospect for electricigenic microorganisms could include any microorganism capable of extracellular electron transfer, even if their capacity for electricity generation has not yet been experimentally evidenced (Lovley et al. 2004). To this respect, hyperthermophiles *Ferroglobus placidus* and *Geoglobus ahangari* were reported to grow at 85 °C by coupling acetate oxidation to Fe(III) reduction (Tor et al. 2001). *Deferribacter thermophilus* isolated from a petroleum reservoir (UK) was able to grow by the reduction of Fe(III) and Mn(IV) and nitrate in the presence of acetate, yeast extract, peptone, and other carbon sources in the temperature range of 50–65 °C (Greene et al. 1997). Kashefi et al. (2003) reported the isolation of a bacterial strain belonging to the *Geobacteraceae* family exhibiting thermophilic growth. The bacterium, *Geothermobacter ehrlichii,* isolated from a hydrothermal vent-coupled acetate oxidation to Fe(III) reduction, with an optimum growth temperature of 55 °C. This strain is the first member in the *Geobacteraceae* family reported to be capable of thermophilic growth. Fe(III) reduction coupled to acetate oxidation has also been demonstrated by the bacterium *Thermincola ferriacetica* (Zavarzina et al. 2007). Overall, a broad range of thermophilic electricigenic microorganisms might be capable of forming a syntrophic consortium with thermophilic carboxydotrophic microorganisms for efficient operation of a syngas-fed MFC.

### 4.1.3.2 Syngas Conversion to Electricity by Thermophilic Hydrogenogenic and Electricigeneic Carboxydotrophs

This pathway leads to CO conversion to $H_2$ through the biological water–gas shift (WGS) reaction (BWGSR) followed by $H_2$ utilization as an electron donor by electricigenic bacteria leading to the generation of $H_2$. It can be defined according to the following stoichiometric equation, which describes the water–gas shift reaction (Maness et al. 2005).

$$CO + H_2O \rightarrow H_2 + CO_2$$

The exposure of the microorganisms to CO leads to $H_2$ production due to the stimulation of a CO-oxidizing and $H_2$-evolving enzymatic system (Singer et al. 2006). In a syngas-fed MFC, this pathway is expected to maximize the CE of syngas transformation, as compared to $H_2$ utilization through acetate formation. A mixed culture of electricigenic and mesophilic carboxydotrophic hydrogenogenic bacteria would allow for CO conversion to $H_2$ and the subsequent use of the $H_2$ produced from CO to $H_2$ present in syngas for electricity generation.

Carboxydotrophic hydrogenogenic microorganisms reported in Sect. 3.1 such as *Thermolithobacter carboxydivorans*, *Carboxydothermus hydrogenoformans*, *T. carboxydophila*, *Carboxydocella thermoautotrophica*, *Themolithobacter carboxydivorans*, and *Carboxydibrachium pacificum* produce $H_2$ from CO oxidation under thermophilic growth conditions (Sokolova et al. 2002, 2005, 2007, 2009). A number of carboxydotrophic microorganisms capable of hydrogenogenic activity have been isolated from marine hydrothermal vents (Henstra et al. 2007b; Sokolova et al. 2009). *C. pacificum*, isolated from a submarine hot vent, grew chemolithotrophically on CO-producing equimolar quantities of $H_2$ and $CO_2$. Its growth was observed between 50 and 80 °C with an optimum temperature of 70 °C (Sokolova et al. 2002). Likewise, *C. thermautotrophica*, a thermophilic CO-utilizing bacterium isolated from a terrestrial hot vent on the Kamchatka Peninsula (Russia) produced $H_2$ and $CO_2$ with a generation time of 1.1 h at a temperature of 58 °C and pH 7. *Carboxydocella sporoproducens*, also isolated from hot springs of Karymshoe Lake, Kamchatka Peninsula (Russia), grows chemolithotrophically on CO (doubling time of 1 h)-producing equimolar quantities of $CO_2$ and $H_2$. The temperature and optimum pH were observed to be 60 °C and 6.8, respectively (Slepova et al. 2006). Sokolova et al. (2005) reported the isolation of the alkali-tolerant carboxydotrophic hydrogenic bacterium, *T. carboxydophila*, from a hot spring of the Baikal Lake region, Russia. CO was found to be the sole source of energy for this bacterium. For lithotrophic growth of *T. carboxydiphila,* acetate or yeast extract was required, but these substrates did not support growth in the absence of CO. Neither acetate nor methanol formation was detected during its growth on CO. Similar to the mesophilic co-culture, a co-culture of thermophilic hydrogenogenic carboxydotrophic microorganisms with $H_2$ utilizing thermophilic electricigenic microorganisms such as *D. thermophilus* or *Pyrobaculum islandicum* could be used for electricity generation in a syngas-fed MFC. *D. thermophilus* was able to grow by the reduction of Fe(III), Mn(IV), and nitrate in the presence of $H_2$. Similarly, *P. islandicum* is able to reduce Fe(III) and Mn(IV) with $H_2$ as an electron donor (Kashefi and Lovley 2000). *Thermolithobacter ferrireducens* reduces Fe(III), anthraquinone-2,6-disulfonate (AQDS), thiosulfate, and fumarate with $H_2$ serving as the electron donor in a temperature range of 50–75 °C (Sokolova et al. 2007).

#### 4.1.3.3 Direct Conversion of CO to Electricity by Thermophilic Iron-Reducing Carboxydotrophs

Most of the metal-reducing carboxydotrophic organisms are thermophiles. High temperatures may be more favorable as less cooling of syngas would be required (Henstra et al. 2007). Although MFC operation at thermophilic temperatures is expected to have a detrimental effect on the CO solubility in the anodic liquid, this is counteracted by the increase in the mass transfer rate with increasing temperature (Drew 1981). Elevated temperatures would also lead to reduced $O_2$ solubility, which is beneficial considering the sensitivity of carboxydotrophs to $O_2$ (Davidova et al. 1994). Iron-reducing carboxydotrophs discussed in Sect. 3.1 include *Thermosinus carboxydovorans*, *Carboxydothermus ferreducens*, *Carboxythermus siderophilus*, *M. perchloratireducens*, and *Thermicola ferriacetica*. All of these produce $H_2$ during the oxidation of CO to $CO_2$. *T. carboxydivorans* grows at temperatures between 40 and 68 °C (with an optimum at 60 °C) at neutrophilic conditions. The bacterium can utilize CO as its sole energy source with a doubling time of 1.15 h leading to the formation of $H_2$ and $CO_2$ in equimolar quantities. Fe (III) was also reduced during its growth on sucrose and lactose. The species was the first metal-reducing carboxydotrophic bacterium to be reported (Sokolova et al. 2004b). *C. ferrireducens* was also isolated from Yellowstone National Park (Henstra and Stams 2004; Sokolova et al. 2009) and has the ability to use CO as an electron donor for AQDS and fumarate reduction. Fumarate, AQDS, ferric iron, and thiosulfate could serve as electron acceptors during its growth on glycerol and $H_2$. *Carboxydothermus siderophilus,* isolated from hot spring of Geyser Valley (Kamchatka Peninsula, Russia), produced $H_2$ and $CO_2$ along with Fe(III) and AQDS reduction during its growth on CO (Slepova et al. 2009). However, the doubling time for *C. siderophilus* (9.3 h) on CO was much longer than that of *T. carboxydivorans* (1.15 h). Balk et al. (2008) reported the isolation of *M. perchloratireducens*, the thermophilic Gram-positive bacterium with the ability to use perchlorate as a terminal electron acceptor. This strain was able to use CO, methanol, pyruvate, glucose, fructose, mannose, xylose, pectin, and cellobiose for its growth. *T. ferriacetica*, isolated from ferric deposits of a terrestrial hydrothermal spring (Kunashir Island, Russia), and is a thermophilic facultative chemolithoautotrophic anaerobic bacterium. It was able to utilize $H_2$ and acetate as energy sources, with Fe(III) serving as the electron acceptor. Also, it was able to grow in an atmosphere of 100 % CO (as the sole energy source), leading to the formation of $H_2$ and $CO_2$. However, it required 0.2 g $L^{-1}$ of acetate as its carbon source during its growth on CO (Zavarzina et al. 2007). *C. hydrogenoformans*, a close relative of *T. ferrireducens,* oxidized CO and $H_2$ using AQDS as an electron acceptor. $CO_2$ and $H_2$ were formed during its growth on CO.

#### 4.1.3.4 Syngas Conversion to Electricity by Thermophilic Methanogenic and Electricigenic Carboxydotrophs

The methanogenic carboxydotrophs identified by Hussain et al. (2012) in the syngas-fed MFC operated at 50 °C are listed in (Table 4.1). The ability to grow in the presence of CO or utilize CO as a sole source of energy has been reported in phylogenetically and physiologically diverse methanogenic groups (see Sect. 3.5). The identified methanogens in Table 4.1 have the capacity for the sequential reduction of $CO_2$ to $CH_4$ in the hydrogenotrophic pathway. This explains the significant amount of $CH_4$ in the anodic off-gas (Hussain et al. 2012). Methanogenic species including *Methanothermobacter wolfeii, Methanothermobacter thermautotrophicus,* and *Methanobrevibacter arboriphilicus* were found (Table 4. 2). These species utilize $H_2$ and $CO_2$ for growth and $CH_4$ formation (Daniels et al. 1977; Winter et al. 1983). In addition, the ability of *M. thermoautotrophicus* and *M. arboriphilicus,* to remove CO in the gas phase while growing on $CO_2$ and $H_2$, has been reported (Daniels et al. 1977). *M. thermoautotrophicus* could utilize CO as the sole energy source by disproportionating CO to $CO_2$ and $CH_4$. This ability could be attributed to the presence of carbon monoxide dehydrogenase (CODH) and acetyl-CoA synthase (ACS) in the microorganisms, the two metalloenzymes fundamental for growth on CO (Oelgeschlager and Rother 2008). The other uncultured archaea identified in the MFC that possess the hydrogenotrophic pathway for $CH_4$ formation belonged to the genera *Methanobacterium* and *Methanobrevibacterium* (Wasserfallen et al. 2000).

**Table 4.2** Methanogenic carboxydotrophs species identified by Hussain et al. (2012) in the syngas-fed MFC operated at 50 °C

| Identified microorganism | Nucleotide identity (%) | Origin/growth conditions | Hypothesized activity in a syngas-fed MFC | References |
|---|---|---|---|---|
| *Methanothermobacter wolfeii* | 100 | $H_2/CO_2$ | Hydrogenotrophic methanogenesis | Winter et al. (1984) |
| *Methanothermobacter thermautotrophicus* | 100 | $H_2/CO_2$ | Hydrogenotrophic methanogenesis | Nolling et al. (1993) |
| Uncultured archaeon | 100 | Isolated from a thermophilic (53 °C) anaerobic biowaste fermenter | – | Malin et al. (2008) |
| *Methanobrevibacter arboriphilicus* | 100 | $H_2/CO_2$ | Hydrogenotrophic methanogenesis | Watanabe et al. (2004) |
| Uncultured archaeon | 100 | Isolated from a microbial digester | – | Wagner et al. (2011) |
| Uncultured *Methanobrevibacterium* | 99 | Isolated from a high temperature natural gas field in Japan | Hydrogenotrophic methanogenesis | Mochimaru et al. (2007) |
| Uncultured *Methanobacterium* | 99 | Isolated under low-hydrogen conditions | Hydrogenotrophic methanogenesis | Sakai et al. (2009) |
| Uncultured *Methanobacterium* | 99 | Isolated from a high temperature petroleum reservoir | Hydrogenotrophic methanogenesis | Kobayashi et al. (2011) |

*nd* not determined

## 4.1.4  Design Considerations of MFCs Operating at Thermophilic Temperatures

The performance and cost of electrodes are important factors affecting the design of MFCs (Wei et al. 2011a). A wide range of electrode materials and configurations have been examined in recent years to improve the performance and reduce cost. A suitable electrode must be a good conductor, chemically stable, mechanically strong, and not expensive (Wei et al. 2011a). Identifying materials and architectures which maximize power generation and CE is a major challenge in designing MFCs (Logan 2008). Another challenge is to reduce cost and develop configurations which can be constructed from a practical point of view (Logan 2008). According to Logan and Regan (2006), the most significant impediment in achieving high power densities in MFCs is the system configurations and not the composition of the bacterial community. Utilizing electrodes with improved properties will enhance the performance of MFCs because different anode materials result in different activation polarization losses (Du et al. 2007). Because the power output of MFCs is low relative to other type of fuel cells, reducing their cost is essential if power generation using this technology is to be an economical method of energy production (Liu and Logan 2004). Many studies have focused on maximizing the power generation in MFCs; however, work on cost minimization studies is limited. Practical applications of MFCs will require developing designs that will not only produce high-power outputs and Coulombic efficiencies but also economical to manufacture in large quantities (Logan 2008).

While the majority of MFCs have been tested at ambient or mesophilic temperatures, thermophilic systems warrant evaluation because of the potential for increased microbial activity rates on the anode. Microbial fuel cell studies at elevated temperatures have been scattered and most use designs that are already established, such as air-cathode single-chamber and two-chamber designs. Previous modular MFC design (Rismani-Yazdi et al. 2007) has shown to work under mesophilic conditions (39 °C), but was not usable at 60 °C. This design was not a closed system and permitted evaporation, specifically from the cathode chamber. As the catholyte evaporated, the anolyte diffused through the proton permeable membrane into the cathode compartment and evaporated. Between 50 and 70 % of the anode, working volume was lost within 2 days. The concentrated anolyte can be detrimental to microbial metabolism and activity due to enrichment of metabolites and cell debris. Thermophilic studies have not addressed these problems other than to note periodic anolyte and catholyte replacement (Mathis et al. 2008; Marshall and May 2009). Jong et al. (2006) utilized continuous flow, rather than batch or fed-batch, which allowed for constant replacement of anolyte and catholyte in the thermophilic MFC. The best MFC performance was with 338 and 11 $cm^3$ $h^{-1}$ for the catholyte and anolyte flow rates, respectively (Jong et al. 2008). The catholyte required a higher flow rate likely due to the continuous evaporation of liquid from the open-cathode chamber. While this prevents drastic liquid loss, electricity production then relies on the electrochemically active biofilm alone since suspended cells are removed with the continuous flow of the anolyte. Several MFC studies have tested a range of operation temperatures and demonstrated consistently higher power densities with higher temperatures, within the limits of the microbial populations (Choi 2004; Moon et al. 2006).

Carver et al. (2011) presented a design based in the original concept elucidated by Min and Angeladaki (2008). The MFC utilized an anaerobic, glass reactor design in combination with a cathode chamber submersed in anolyte. Rather than having an extensive layers of gaskets, membrane, carbon paper, and polycarbonate as in the previous design (Min and Angeladaki 2008), the cathode chamber had a single-rubber O-ring that is able to prevent liquid or air crossover. The components of the cathode assembly, including the stainless screws, foil and graphite disks, have all been shown to be conducive and were securely connected. Analyses of the glucose-fed thermophilic MFC showed improved performance over 120 h with an increased maximum power of 3.3–4.5 mW m$^{-2}$ (Carver et al. 2011). The polarization curve has three distinct sections of irreversible voltage losses: activation loss, ohmic loss, and mass transfer loss. The typical initial and drastic voltage drop was not apparent, indicating lower than normal activation losses (Carver et al. 2011). This is attributed to increased reaction rates at thermophilic temperatures that lowered the activation energy and therefore the voltage necessary to maintain active, anaerobic metabolism. Ohmic loss can be observed in the center of the polarization curve with the gradual decrease of voltage as current density increases (Carver et al. 2011). The slope of this overpotential section, equivalent to voltage over current, yielded an internal resistance of $9.25 \pm 0.15$ $\Omega$. This value is in the general range reported for other MFCs, although the experimental conditions are not comparable among the studies reviewed in the literature (He et al. 2006; Ieropoulos et al. 2010). The results suggest the potential for stable, thermophilic MFC operation although optimization of biological and engineering components is necessary prior to application of the design.

A CO/syngas-fed MFC system requires a CO-tolerant cathode. The most extensively used cathode material in a conventional MFC consists of carbon paper with a Pt/C catalyst (Logan 2008). Even though the Pt-based cathode demonstrates high electrochemical activity, the use of Pt is undesirable due to high costs and easy inhibition by CO (Logan 2008; Herrmann et al. 2009). Even at small concentrations, CO can fully cover the Pt surface, thereby reducing the reaction site. CO is easily able to absorb to Pt due to the negative free energy of adsorption (Baschuk and Li 2001). Mehta et al. (2010) used a CoTMPP cathode to generate electricity from CO with a Co load of 0.5 mg cm$^{-2}$. A maximum power density of 6.4 mW L$^{-1}$ was reported. The cathode performance was tested in acetate- and CO-fed MFCs. MFC operation on CO showed the best performance with the CoTMPP/FeTMPP/C cathode catalyst. Considering the high cost of Pt-based cathodes and the plausible decrease in activity with time, the use of CoTMPP/FeTMPP/C or FePc cathodes is a step forward in increasing the efficiency of CO-operated MFCs.

Membrane systems also need to be considered for improved mass transfer efficiency. The primary resistance to gas transport is in the liquid film of the gas–liquid interface. To improve transfer efficiency, the conventional continuous stirred tank reactors (STRs) have been used. However, the high impeller speeds require a high-power input (Henstra et al. 2007b; Hickey et al. 2008) and leads to biofilm shearing, causing a decrease in the growth of shear-sensitive microorganisms (Munasinghe and Khanal 2010). As a solution, a bubble-free gas transfer to liquid has been accomplished by the selection of a membrane system with a high selectivity for the gaseous substrate. The membrane systems offer an efficient and a relatively inexpensive method for gas–liquid mass transfer (Scott and Hughes 1996). Silicone membranes have also been used. These are dense membranes, which offer the advantage of high mechanical strength, flexibility, and stability under high temperature and pressures. They have been reported to be ideal for membrane-based bubble-less aeration without vigorous mixing, where a conventional system is unable to meet the $O_2$ requirements of a high-rate system (Côté et al. 1989). Other promising alternatives to the conventional STRs for increased gas–liquid mass transfer include monolith packing and columnar reactors. Monolith packing consists of a number of narrow, straight, and parallel flow channels with a large-open frontal area which allows for a low flow resistance, leading to low pressure drops and low energy losses. High volumetric mass transfer rates of $\sim 1$ $s^{-1}$ and a 50–80 % reduction in power consumption as compared to conventional reactors make monolith reactors an economically viable option (Hickey et al. 2008; Munasinghe and Khanal 2010). Similarly, columnar reactors such as bubble column, trickle bed, and airlift reactors offer the advantage of a high gas–liquid mass transfer rate with low operational and maintenance costs. $K_L$ values within the range of 18–860 $h^{-1}$ have been reported for such reactors (Charpentier 1981; Bredwell et al. 1999; Munasinghe and Khanal 2010).

Several reactor design improvements such as the low-frequency vibration of liquid phase in a bubble column reactor, the addition of static mixers, baffles, perforated plates, jet loop, and forced circulation loop in internal- and external-loop airlift reactors promise a further increase in the gas–liquid mass transfer efficiency (Chisti et al. 1990; Gavrilescu et al. 1997; Vorapongsathorn et al. 2001; Krichnavaruk and Pavasant 2002; Ugwu and Ogbonna 2002; Ellenberger and Krishna 2003; Fadavi and Chisti 2005).

## 4.2 Biofuel and Organic Acid Production by Thermophilic Carboxydotrophs from Synthesis Gas (Syngas) Fermentation

A significant portion of biomass sources such as straw and wood is poorly degradable and cannot be converted to biofuels by microorganisms. Biomass consists of cellulose, hemicellulose, and lignin, and the latter of which is extremely resistant to degradation. One approach to unlocking the potential in this abundant feedstock is to separate the lignin from the carbohydrate fraction of the biomass via extensive pretreatment of the lignocellulose involving, for example, steam-explosion and/or acid hydrolysis. These pre-treatments are designed to allow the carbohydrate portion of the biomass to be broken down into simple sugars, for example, by enzymatic hydrolysis using exogenously added cellulases to release fermentable sugars (Carere et al. 2008). Such approaches have been found to be expensive and rate-limiting (Datar et al. 2004; Carere et al. 2008; Köpke et al. 2011). Alternatively, processes using cellulolytic microorganisms (*Cl. cellulolyticum, C. thermocellum,* and *C. phytofermentans*) to carry out both the hydrolysis of lignocelluloses and sugar fermentation in a single step, termed "consolidated bioprocessing process (CBP)" (Carere et al. 2008; Plecha et al. 2013), have been proposed; however, the development of these is still at an early stage, and again low conversion rates seem to be a major limitation that needs to be overcome.

The gasification of this waste material to produce synthesis gas could offer a solution to this problem, as microorganisms that convert CO and $H_2$ (the essential components of syngas) to multi-carbons are available (Table 4.3). Due to the flexibility of the microbes to ferment syngas with diverse composition, virtually any carbonaceous materials can be used as feedstock for gasification. Non-food biomass that can be employed as feedstock for gasification includes agricultural wastes, dedicated energy crops, forest residues, and municipal organic wastes, or even glycerol and feathers (Abubackar et al. 2011; Mohammadi et al. 2011; Siedlecki et al. 2011; Wei et al. 2011b; Dudynski et al. 2012). Biomass is available on a renewable basis, either through natural processes or anthropogenic activities (e.g., organic wastes). It has been estimated that out of a global energy potential from modern biomass of 250 EJ per year in 2005, only 9 EJ (3.6 %) was used for energy generation (Siedlecki et al. 2011). The use of existing waste streams such as municipal organic waste also differentiates itself from other feedstocks such as dedicated energy crops because these wastes are available today at economically attractive prices, and they are often already aggregated and require less indirect land use. Alternatively, gasification of non-biomass sources such as coal, cokes, oil shale, tar sands, sewage sludge, and heavy residues from oil refining, as well as reformed natural gas, are commonly applied as feedstocks for the FTP and can also be used for syngas fermentation (Klasson et al. 1992; Sipma et al. 2006). Furthermore, some industries such as steel manufacturing, oil refining, and chemical production generate large volume of CO- and/or $CO_2$- rich gas streams as wastes. Tapping into these sources using microbial fermentation process essentially converts existing toxic waste gas streams into valuable commodities such as biofuels. The overall process of gas fermentation is outlined in Fig. 4.11. Prior to gasification, biomass generally needs to go through a pre-treatment process encompassing drying, size reduction (e.g., chipping, grinding, and chopping), pyrolysis, fractionation, and leaching depending on the gasifier configuration (Griffin et al. 2012; McKendry 2002). This upstream pre-treatment process can incur significant capital expense and add to the overall biomass feedstock cost, ranging from US$16–70 per dry ton (Griffin et al. 2012).

Gasification is a well-established technology where carbonaceous material is cracked at extreme temperatures (700–1,000 °C) (Mohammadi et al. 2011; Griffin et al. 2012). If pure oxygen is used as the oxidant in the gasifier, then the resulting

**Table 4.3** Anaerobic carboxydotrophic microorganisms

| Organism | Substrate | Products | $T_{opt}$ (°C) | $pH_{opt}$ | References |
|---|---|---|---|---|---|
| *Mesophilic microorganisms* | | | | | |
| *Acetobacterium woodii* | $H_2/CO_2$, CO | Acetate | 30 | 6.8 | Genthner and Bryant (1987), Poehlein et al. (2012) |
| *Acetonema longum* | $H_2/CO_2$ | Acetate, butyrate | 30–33 | 7.8 | Kane and Breznak (1991) |
| *Alkalibaculum bacchi* | $H_2/CO_2$, CO | Acetate, ethanol | 37 | 8.0–8.5 | Allen et al. (2010), Liu et al. (2012) |
| *Blautia producta* | $H_2/CO_2$, CO | Acetate | 37 | 7.0 | Lorowitz and Bryant (1984) |
| *Butyribacterium methylotrophicum* | $H_2/CO_2$, CO | Acetate, ethanol, butyrate, butanol | 37 | 6.0 | Lynd et al. (1982), Zeikus et al. (1980), Grethlein et al. (1991) |
| *Citrobacter* sp. Y19 | $H_2/CO_2$, CO | $H_2$ | 30–40 | 5.5–7.5 | Jung et al. (1999b, 2002) |
| *Clostridium aceticum* | $H_2/CO_2$, CO | Acetate | 30 | 8.3 | Lux and Drake (1992), Adamse et al. (1980), Braun et al. (1981) |
| *Clostridium autoethanogenum* | $H_2/CO_2$, CO | Acetate, ethanol, 2,3-butane-diol, lactate | 37 | 5.8–6.0 | Abrini et al. (1994), Köpke et al. (2011) |
| *Clostridium carboxi-divorans* "P7" | $H_2/CO_2$, CO | Acetate, ethanol, butyrate, butanol, lactate | 38 | 6.2 | Bruant et al. (2010), Liou et al. (2005) |
| *Clostridium drakei* | $H_2/CO_2$, CO | Acetate, ethanol, butyrate | 25–30 | 5.8–6.9 | Gössner et al. (2008), Küsel et al. (2000), Liou et al. (2005) |
| *Clostridium formicoaceticum* | CO | Acetate, formate | 37 | NR | Andreese et al. (1970), Diekert and Thauer (1978), Lux et al. (1992) |
| *Clostridium glycolicum* | $H_2/CO_2$ | Acetate | 37–40 | 7.0–7.5 | Ohwaki et al. (1977) |

(continued)

**Table 4.3**   (continued)

| Organism | Substrate | Products | $T_{opt}$ (°C) | $pH_{opt}$ | References |
|---|---|---|---|---|---|
| *Clostridium ljungdahlii* | $H_2/CO_2$, CO | Acetate, ethanol, 2,3-butanediol, lactate | 37 | 6.0 | Köpke et al. (2011, 2010), Tanner et al. (1993) |
| *Clostridium magnum* | $H_2/CO_2$ | Acetate | 30–32 | 7.0 | Bomar et al. (1991), Schink (1984) |
| *Clostridium mayombei* | $H_2/CO_2$ | Acetate | 33 | 7.3 | Kane et al. (1991) |
| *Clostridium methoxybenzovorans* | $H_2/CO_2$ | Acetate, formate | 37 | 7.4 | Mechichi et al. (1999) |
| *Clostridium ragsdalei* | $H_2/CO_2$, CO | Acetate, ethanol, 2,3-butanediol, lactate | 37 | 6.3 | Huhnke et al. (2008) |
| *Clostridium scatologenes* | $H_2/CO_2$, CO | Acetate, ethanol butyrate | 37–40 | 5.4–7.5 | Gössner et al. (2008), Küsel et al. (2000) |
| *Eubacterium limosum* | $H_2/CO_2$, CO | Acetate | 38–39 | 7.0–7.2 | Genthner and Bryant (1987), Genthner et al. (1981) |
| *Oxobacter pfennigii* | $H_2/CO_2$, CO | Acetate, butyrate | 36–38 | 7.3 | Krumholz and Bryant (1985) |
| *Peptostreptococcus productus* | $H_2/CO_2$, CO | $H_2$ | 37 | 7.0 | Lorowitz and Bryant (1984) |
| *Rubrivivax gelatinosus* | $H_2/CO_2$, CO | $H_2$ | 34 | 6.7–6.9 | Dashekvicz and Uffen (1979); Uffen (1976) |
| *Rhodopseudomonas palustris* P4 | $H_2/CO_2$, CO | $H_2$ | 30 | NR | Jung et al. (1999a) |
| *Rhodospirillum rubrum* | $H_2/CO_2$, CO | $H_2$ | 30 | 6.8 | Kerby et al. (1995) |
| *Thermophilic microorganisms* | | | | | |
| *Archaeoglobus fulgidus* | $H_2/CO_2$, CO | Acetate, formate, $H_2S$ | 83 | 6.4 | Zellner et al. (1989) |
| *Moorella stamsii* | CO | Acetate, $H_2$ | 65 | 7.5 | Alves et al. (2013) |
| *Moorella thermoacetica* | $H_2/CO_2$, CO | Acetate | 55 | 6.5–6.8 | Kerby and Zeikus (1983), Daniel et al. (1990), Pierce et al. (2008) |

(continued)

**Table 4.3** (continued)

| Organism | Substrate | Products | $T_{opt}$ (°C) | $pH_{opt}$ | References |
|---|---|---|---|---|---|
| *Moorella thermoautotrophica* | $H_2/CO_2$, CO | Acetate | 58 | 6.1 | Savage et al. (1987) |
| *Calderihabitans maritimus* | CO | $H_2$ | 65 | 7.0–7.5 | Yoneda et al. (2013) |
| *Carboxydothermus hydrogenoformans* | $H_2/CO_2$, CO | $H_2$ | 70-72 | 6.8-7.0 | Svetlitchnyi et al. (2001) |
| *Carboxydibrachium pacificus* | $H_2/CO_2$, CO | $H_2$ | 70 | 6.8–7.1 | Rother and Metcalf (2004) |
| *Carboxydocella sporoproducens* | $H_2/CO_2$, CO | $H_2$ | 60 | 6.8 | Slepova et al. (2006) |
| *Carboxydocella thermoautotrophica* | $H_2/CO_2$, CO | $H_2$ | 58 | 7.0 | Sokolova et al. (2002) |
| *Desulfotomaculum kuznetsovii* | $H_2/CO_2$, CO | Acetate, $H_2$, $H_2S$ | 60 | 7.0 | Parshina et al. (2005b) |
| *Desulfotomaculum thermobenzoicum subsp. thermosyntrophicum* | $H_2/CO_2$, CO | Acetate, $H_2$, $H_2S$ | 55 | 7 | Parshina et al. (2005a) |
| *Desulfotomaculum carboxydivorans* | $H_2/CO_2$, CO | $H_2$, $H_2S$ | 55 | 7.0 | Parshina et al. (2005b) |
| *Methanothermobacter thermoautotrophicus* | CO | $CH_4$ | 6.8 | NR | Daniels et al. (1977) |
| *Thermincola carboxydiphila* | $H_2/CO_2$, CO | $H_2$ | 55 | 8.0 | Sokolova et al. (2005) |
| *Thermincola ferriacetica* | $H_2/CO_2$, CO | $H_2$ | 57–60 | 7.0–7.2 | Zavarzina et al. (2007) |
| *Thermoanaerobacter kiuvi* | $H_2/CO_2$ | Acetate | 66 | 6.4 | Leigh and Wolfe (1983) |
| *Thermococcus* strain AM4 | CO | $H_2$, $H_2S$ | 82 | 6.8 | Sokolova et al. (2004a) |
| *Thermolithobacter carboxydivorans* | $H_2/CO_2$, CO | $H_2$ | 70 | 7.0 | Svetlichnyi et al. (1994), Sokolova et al. (2007) |
| *Thermosinus carboxydivorans* | $H_2/CO_2$, CO | $H_2$ | 60 | 6.8–7.0 | Sokolova et al. (2004b) |

*NR* Not reported

synthesis gas is rich in CO and $H_2$. If air is used, then the resulting gas (producer gas) is a mixture of CO, $CO_2$, $H_2$, $CH_4$, $N_2$, and some light hydrocarbons such as $C_2H_2$ and $C_2H_4$ as well as heavy hydrocarbons known as tars (Munasinghe and Khanal 2010; Abubackar et al. 2011; Griffin et al. 2012). The partial oxidation reactions with oxygen that take place within a gasifier are exothermic. Steam can also be used as an oxidant in indirect gasification. The result of these

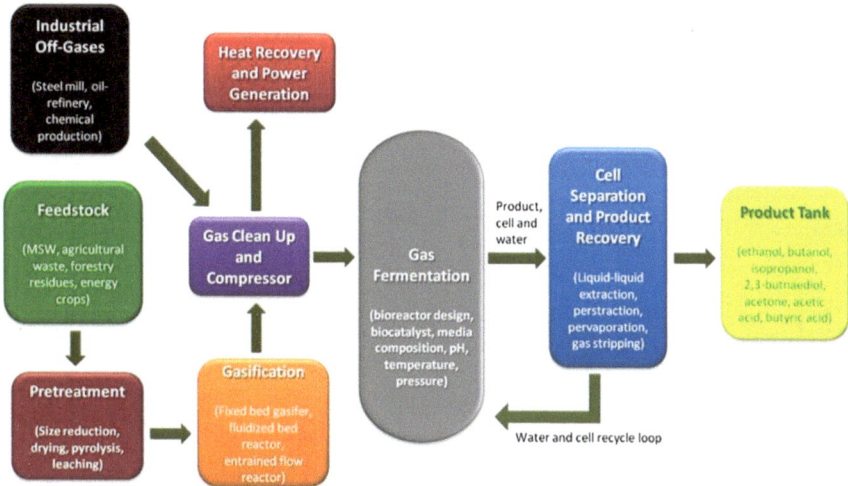

**Fig. 4.11** Overview of gas fermentation process (Liew et al. 2013)

thermochemical reactions is an endothermic and often heat transfer limited, but thermodynamically efficient process (Sipma et al. 2006). The ratio of the components of synthesis gas varies depending on the biomass source and the gasification conditions employed (see Table 4.4). Gasification of biomass to provide vehicles with fuel has been in use since the early 1930s (McKendry 2002). Due to petroleum products shortage during World War II, this technology flourished in some European countries providing fuel for both civilians and militaries (Dasappa et al. 2004). More recently, in the 1980s and 1990s, synthesis gas was successfully used in the USA and Europe for heat and electricity (Faaij 2006).

**Table 4.4** Composition of synthesis gas derived from various lignocellulosic biomass sources

| Source | Composition (vol. %) | | | | | | References |
|---|---|---|---|---|---|---|---|
| | CO | $CO_2$ | $H_2$ | $N_2$ | $CH_4$ | Other | |
| Switchgrass | 14.7 | 16.5 | 4.4 | 56.8 | 4.2 | 3.4 | Datar et al. (2004) |
| Pine wood chips | 16.1 | 13.6 | 16.6 | 37.6 | 2.7 | 13.4 | Corella et al. (1998) |
| Willow | 9.4 | 17.2 | 7.2 | 60.5 | 3.3 | 2.5 | van der Drift et al. (2001) |
| Cacao shells | 8 | 16 | 9 | 61.5 | 2.3 | 3.2 | van der Drift et al. (2001) |
| Dairy biomass | 8.7 | 15.7 | 18.6 | 56 | 0.6 | 0.4 | Gordillo and Annamalai (2010) |
| Kentucky blue-grass straw | 12.9 | 17.4 | 2.6 | 64.2 | 2.1 | 0.8 | Boteng et al. (2007) |
| Demolition wood/paper residue | 9.2 | 16.1 | 6.1 | 63.2 | 2.8 | 2.6 | van der Drift et al. (2001) |

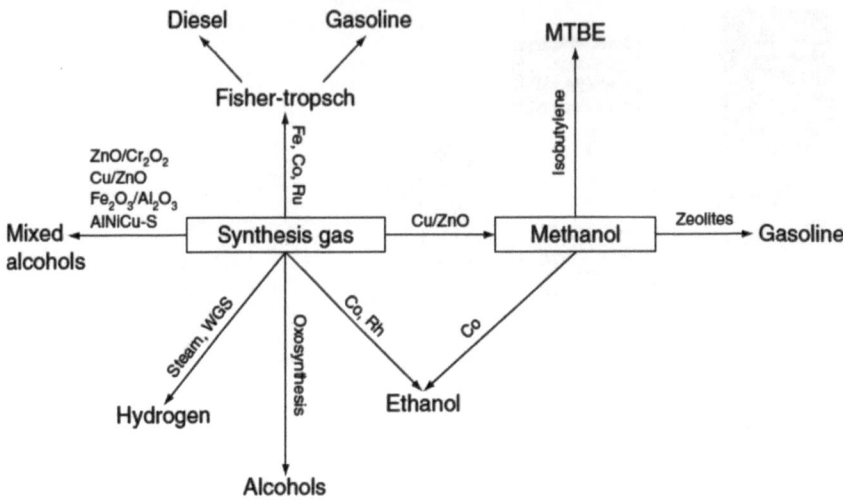

**Fig. 4.12** Fuel products obtained from synthesis gas transformation (Spath and Dayton 2003)

Synthesis gas can also be transformed into a number of different products such as methanol, ethanol, hydrogen, dimethylether, and others via chemical catalysts (Fig. 4.12). This idea is not new and has been developed in the past few decades (Wilhelm et al. 2001). Nevertheless, these are expensive processes subjected to high pressures and temperatures (Takeguchi et al. 2000; Quinn et al. 2004). Furthermore, syngas coming out of the gasifier has many contaminants leading to catalyst poisoning (Leibold et al. 2008). Some synthesis gas transformations to biofuels such as ethanol, butanol, and hydrogen can be performed using chemical as well as biological catalysts (Fig. 4.13a, b). Biological processes, while slower than chemical reactions, have a number of advantages such as higher yields (even similar to direct fermentation of biomass) (Spath and Dayton 2003), specificity, and generally greater potential for decreased catalyst poisoning (Younesi et al. 2008). The thermochemical conversion of lignocellulosic biomass integrates the process of biomass gasification and biofuel synthesis as illustrated in Fig. 4.14. Production of biofuel from syngas is either performed using inorganic or metal-based catalysts known as Fischer–Tropsch (FT) process or microbial catalysts known as syngas fermentation (Mohammadi et al. 2011).

The CO and $H_2$ present in syngas are substrates for microbial metabolism, which can be exploited for the synthesis of various interesting products. It is expected that syngas fermentation will play a role in the conversion of biomass, wastes, and residues that form poor substrates for direct fermentation. As gasification results in gas with a high temperature, thermophilic microbial processes might be most applicable for the biotechnological production of chemicals from syngas.

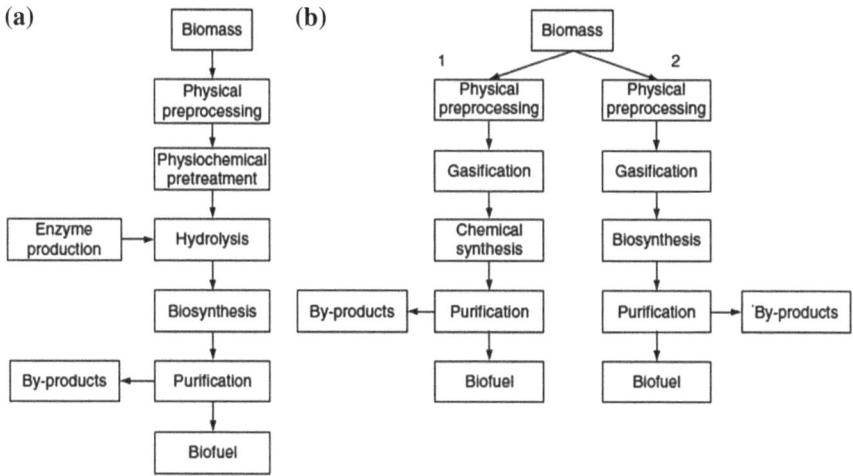

**Fig. 4.13** Biofuel production from biomass: (**a**) hydrolysis–fermentation (**b1**) gasification–chemical synthesis, and (**b2**) gasification–biosynthesis (Tirado-Acevedo et al. 2010)

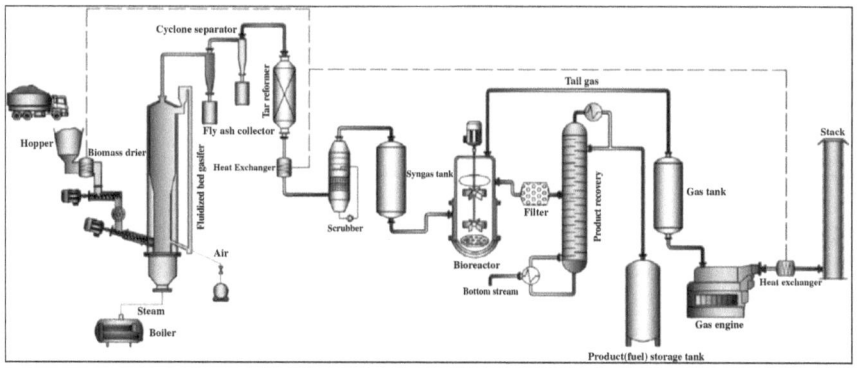

**Fig. 4.14** Schematic representation of biomass gasification integrated with syngas fermentation (Mohammadi et al. 2011)

## 4.2.1 Microbiology of Carboxydotrophic Microorganisms Capable of Converting Syngas to Biofuels and Organic Acids

Microorganisms such as carboxydotrophic acetogens, carboxydotrophic hydrogenogens, carboxydotrophic methanogens, carboxydotrophic sulfate-reducing bacteria, and carboxydotrophic iron-reducing bacteria are able to utilize the $CO_2 + H_2$ and/or CO (see Chap. 3) available in such syngas as their sole source of carbon and energy for growth as well as the production of biofuels and other valuable products.

However, the majority of carboxydotrophs that is described to synthesize metabolic end products that have potentials as liquid transportation fuels so far are the acetogenic carboxydotrophs. Production of simple organic compounds from syngas is thermodynamically favorable, as shown by the negative changes of Gibbs free energy, $\Delta G°$, in Table 4.1. Fisher and Tropsch were the first to demonstrate that acetate could be produced from CO by anaerobic sewage sludge in 1932 (Diekert et al. 1982). Since then, microbial fermentation of syngas produced from cellulosic materials has been studied by many and has even been demonstrated in full-scale operation. Although difficulties had been encountered in early attempts, many now believe that microbial production of fuels and commercial products from syngas is commercially feasible through proper product formation, microbe selection, and process optimization (Barik et al. 1990). Many anaerobic microorganisms (some have been isolated but not yet identified) were shown capable of growth with CO and $H_2$ as substrates (Table 4.3). Although most strains showed the formation of acetate, formate, butyrate, ethanol, and butanol were also reported as products. Additionally, several purple non-sulfur bacteria were isolated that are able to convert CO to $H_2$ in a process similar to the WGS reaction (Table 4.3).

One of the most significant factors of syngas fermentation is the CO conversion rate and superior specific products of microorganisms which could produce valuable fuels and organic compounds. Isolation of new microbes capable of converting CO into fuels and organics and research to improve efficiency will make this technology more commercially feasible.

### 4.2.1.1  Carboxydotrophs that Produce Acetate from CO or Syngas

Many early studies were conducted on microorganisms that produce acetate and butyrate using CO as substrate. These include the homoacetogens (acetogenic microorganisms that produce acetate as major fermentation product) (Table 4.3).

Mesophilic Carboxydotrophs that Produce Acetate

*Eubaterium limosum* isolated from various environments, e.g., human intestine, sewage, rumen, and soil, produces acetate and butyrate exclusively using CO as the sole energy source (Genthner et al. 1982, 1987). Other homoacetogens, including *Clostridium aceticum* (Wieringa 1936, 1939) and *Acetobacterium woodii* (Balch et al. 1977), were identified to be able to produce acetate from syngas gas. *C. aceticum* was the first acetogen to be isolated, from a soil sample in 1936, although the strain was subsequently lost (Braun et al. 1981). In 1980, spores of the original strain were serendipitously found and reactivated (Braun et al. 1981), while at the same time, it was separately re-isolated (Adamse 1980). Acetate is produced from growth-supporting substrates including $H_2$, $CO_2$, CO, and a range of sugars (fructose, ribose, glutamate, fumarate, malate, and pyruvate). *C. aceticum* has an optimal growth temperature of 30 °C (Braun et al. 1981). A genome sequence for the organism is currently under construction (Schiel-Bengelsdorf and Dürre 2012). *C. aceticum* has recently been investigated for the production of acetic acid from synthesis gas (Sim et al. 2007), with a published productivity of 1.28 g $L^{-1}$ with 100 % CO conversion (Sim and Kamaruddin 2008). *A. woodii* is also one of the carboxydotrophic mesophiles capable of producing acetate from CO. *A. woodii* was isolated from estuary sediment in 1977 (Balch et al. 1977) and is capable of forming acetate from $H_2$ to $CO_2$ and CO. With fructose, small amounts of ethanol have been reported under certain conditions (Buschhorn et al. 1989). *A. woodii* has a reported optimal growth temperature of 30 °C and grows on other substrates such glucose, lactate, glycerate, and formate (Balch et al. 1977). Recently, the genome of *A. woodii* was sequenced (Poehlein et al. 2012). Most recently published research on *A. woodii* has focused on elucidating the acetogenic mode of energy conservation (Imkamp et al. 2007; Müller et al. 2008; Biegel et al. 2009; Schmidt et al. 2009) and its sodium-dependant metabolism (Schmidt et al. 2009; Poehlein et al. 2012). However, studies have demonstrated its use as a biocatalyst for the production of acetate, reporting an acetate concentration of 44 g $L^{-1}$ after 11 days, with a maximum cell-specific acetate productivity of 6.9 $g_{acetate}$ $g_{cdw}^{-1}$ $d^{-1}$ achieved with $H_2$ and $CO_2$ as substrate (Demler and Weuster-Botz 2011).

Thermophilic Carboxydotrophs that Produce Acetate

Thermophilic fermentation has the advantages of less cooling requirement of syngas, high conversion rates, and a separation benefit for alcohols; however, higher temperatures would reduce syngas solubility (Henstra et al. 2007b). Gram-positive thermophiles that can convert CO and $H_2O$ into $H_2$ and $CO_2$ also have been isolated recently, including *C. hydrogenoformans* and *T. carboxydivorans*. Other thermophiles, such as *M. thermoautotrophica* and *M. thermoacetica*, have been reported to use CO-producing acetate (Table 4.3).

   *Moorella thermoacetica* is considered a model acetogen (Drake and Daniel 2004) and it was used by Wood and Ljungdahl (Wood 1991) and later by Ragsdale and Pierce (2008) to elucidate the Wood–Ljungdahl pathway. *M. thermoacetica* is a thermophilic bacterium isolated from horse manure and was originally characterized in 1942 as *C. thermoaceticum* (Fontaine et al. 1942). It forms acetate from a diverse range of substrates, with an optimal growth temperature of 55–60 °C. Over the past 10 years, new *Moorella* strains have been isolated and explored for both acetate and ethanol production, although only very low ethanol productivities have been reported (Balk et al. 2003; Sakai et al. 2004; Jiang et al. 2009). Due to its role in the elucidation of the Wood–Ljungdahl pathway, *M. thermoacetica* is well characterized (Drake and Daniel 2004). In 2008, *M. thermoacetica* became the first acetogen to have its genome sequence published (Pierce et al. 2008), helping to improve understanding of the acetogenic metabolism and mode of energy conservation. Using a dilution-cycle fermentation mode, up to 108 g $L^{-1}$ acetate has been produced from sugar using *M. thermoacetica* (Wiegel et al. 1991). Production rates of up to 14.3 g $L^{-1}$ $h^{-1}$ acetate from sugar have been reported in continuous culture, but only at low concentrations of 7.1 g $L^{-1}$ (Reed and Bogdan 1985).

*Archaeoglobus fulgidus* is a strict anaerobic hyperthermophilic archaeon that oxidizes lactate completely to $CO_2$ with sulfate as electron acceptor (Stetter 1988). The genome sequence of *A. fulgidus* encodes three CO-dehydrogenase (CODH) genes (Klenk et al. 1997). *A. fulgidus* grows successfully into growth medium containing sulfate and CO (Henstra et al. 2007a). In the presence of CO and sulfate, the culture $OD_{660}$ increased to 0.41, and sulfide, carbon dioxide, acetate, and formate were formed. Accumulation of formate was transient. Similar results, except that no sulfide was formed, were obtained when sulfate was omitted. Hydrogen was never detected. Under the conditions tested, the observed concentrations of acetate (18 mM) and formate (8.2 mM) were highest in cultures without sulfate. Proton NMR spectroscopy indicated that $CO_2$, and not CO, is the precursor of formate and the methyl group of acetate. Methylviologen-dependent formate dehydrogenase (FDH) activity (1.4 m mol formate oxidized $min^{-1}$ $mg^{-1}$) was detected in cell-free extracts and expected to have a role in formate reuptake. It is speculated that formate formation proceeds through hydrolysis of formyl-methanofuran or formyltetrahydromethanopterin. *A. fulgidus* can grow chemolithoautotrophically with CO as acetogen and is not strictly dependent on the presence of sulfate, thiosulfate, or other sulfur compounds as electron acceptor (Henstra et al. 2007a).

*Moorella thermoautotrophica,* previously known as *C. thermoautotrophicum* is a thermophilic acetogen capable of growth at the expense of a variety of heterotrophic and autotrophic substrates in an undefined, complex medium (Wiegel et al. 1981). On the basis of a nearly 50 % DNA homology, Wiegel et al. (1981) demonstrated that *M. thermoautotrophica* is similar to another thermophilic acetogen, *M. thermoautotrophica*; a subsequent enzyme study demonstrated that these two acetogenic clostridia also display similar enzyme profiles (Clark et al. 1982). Lundie and Drake (Lundie and Drake 1984) defined the nutritional requirements of *M. thermoautotrophica* and demonstrated that glucose could serve as the sole source of carbon and energy and that nicotinic acid was the sole essential vitamin. *M. thermoautotrophica* was adapted to minimal medium and cultivated at the expense of glucose, methanol, or $H_2$–$CO_2$ (Savage and Drake 1986). No supplemental amino acids were required for growth of the adapted strain, and nicotinic acid was the sole essential vitamin. Neither $N_2$ nor nitrate could replace ammonium as the nitrogen source, and biotin was preferentially stimulatory for glucose cell lines. Growth in minimal medium yielded substantially higher acetate concentrations per unit of biomass formed than did growth in undefined medium. Additions of $CO_2$ and possibly $H_2$ are stimulatory to CO-dependent growth of *M. thermoautotrophica* (Savage and Drake 1986; Savage et al. 1987).

Another *Moorella* strain that produce acetate is *Moorella stamsii*. This strain was isolated from anaerobic sludge from a municipal solid waste digester (Alves et al. 2013). The temperature range for growth is between 50 and 70 °C, with an optimum at 65 °C. The pH range for growth is between 5.7 and 8.0, with an optimum at 7.5. *M. stamsii* has the ability to ferment various sugars, such as fructose, galactose, glucose, mannose, raffinose, ribose, sucrose, and xylose, producing mainly $H_2$ and acetate. In addition, the isolate was able to grow with CO as the sole carbon and energy source. CO oxidation was coupled to $H_2$ and $CO_2$ formation. The G + C content of the genomic DNA was 54.6 mol%. Based on 16S rRNA gene sequence analysis, this bacterium is most closely related to *Moorella glycerini* (97 % sequence identity) (Alves et al. 2013).

*Desulfotomaculum kuznetsovii* and *D. thermobenzoicum* subsp. thermophilic carboxydotrophic sulfate reducers are able to grow on CO as the only electron donor and, in particular in the presence of hydrogen/carbon dioxide, at CO concentrations as high as 50–70 % (Pashina et al. 2005b). The latter $SO_4$ reducers coupled CO oxidation to $SO_4$ reduction, but a large part of the CO was converted to acetate. The co-cultures of *C. hydrogenoformans* and *D. kuznetsovii* or *D. thermobenzoicum* subsp. *thermosyntrophicum*, grown with 100 % CO as sole carbon and energy source in standing cultures, convert CO and reduce $SO_4$ (Pashina et al. 2005b). When *C. hydrogenoformans* is cultivated with *D. kuznetsovii* without shaking, $H_2$ is formed gradually, and it is also consumed gradually. At the end of the experiment, 4.3 mM acetate is formed. Under shaken conditions, $H_2$ accumulates rapidly, and its further conversion occurs slowly. Sulfate reduction is inhibited (only 0.4 mM $H_2S$ is formed), and $H_2$ is not consumed completely. More acetate (6.6 mM) is formed compared with the standing cultures. When *C. hydrogenoformans* is cultivated with *D. thermobenzoicum* subsp. *thermosyntrophicum* in standing cultures, the rates of $H_2$ formation and $SO_4$ reduction are similar to the rates in the co-culture with *D. kuznetsovii* (Pashina et al. 2005b). Under shaken conditions, $H_2$ is formed fast, but only after all CO is converted, the $H_2$ concentration decreased and $SO_4$ is reduced (Pashina et al. 2005b). Acetate concentration is 4 mM in standing cultures and 7.5 mM in shaken cultures (Pashina et al. 2005b).

*Thermoanaerobacter kivui* (formerly *A. kivui*) is a thermophilic, anaerobic, non-spore-forming species of bacteria (Leigh and Wolfe 1983). *T. kivui* was originally isolated from Lake Kivu in Africa. The growth range for the organism is 50–72 °C at pH 5.3–7.3, with optimal growth conditions at 66 °C and pH 6.4. The original genus *Acetogenium* was named because organism principally produces acetic acid from substrates (Leigh et al. 1981). Further 16S ribosomal RNA studies put the bacteria into genus *Thermoanaerobacter.*

### 4.2.1.2  Carboxydotrophs that Produce Ethanol from CO or Syngas

Mesophilic Carboxydotrophs that Produce Ethanol

The first microorganism shown to catalyze conversion of synthesis gas components to ethanol was the acetogen *Clostridium ljungdahlii* (Barik et al. 1988). Although ethanol production from synthesis gas was detected, the main product was acetate. Initially, a molar ratio of ethanol-to-acetate of 1:9 and an ethanol concentration of less than 1 g $L^{-1}$ were obtained in batch cultures (Vega et al. 1989). Klasson et al. (1991) noted that yeast extract had an influence on the product ratio and that the ethanol production. Therefore, yeast extract concentration was greatly reduced or eliminated completely and replaced by cellobiose. This increased both ethanol and cell concentrations. Adding reducing agents to the media seemed to alter electron flow to NADH formation and, in turn, increased ethanol production. These first experiments were done in batch cultures. By applying the culture performance information acquired through experimentation and operating two continuously stirred tank reactors (CSTR) in series (the first to promote growth and the second one for increased ethanol production), they were able to improve ethanol production by 30-fold (Klasson et al. 1991). A cell recycle apparatus was added to the CSTR, pH was held at 4.5, agitation was set at 450 rpm, gas flow rate was 30 ml min$^{-1}$, and liquid flow rate ranged from 3.5 to 12 ml h$^{-1}$. These modifications increased the cell concentration from 800 to 4,000 mg L$^{-1}$ and increased ethanol production to 50 g L$^{-1}$ with an ethanol-to-acetate molar ratio of 21:1; and CO and H$_2$ conversions of 90 and 70 %, respectively (Klasson et al. 1991; Phillips et al. 1993). Investigations also showed that *C. ljungdahlii* is quite tolerant of sulfur gases. It is able to grow and uptake CO and H$_2$ in the presence of up to 2.7 % H$_2$S or 5 % carbonyl sulfide (Klasson et al. 1993; Smith et al. 1991). This is relevant since syngas contains a considerable amount of these gases. This organism favors the production of acetate during its active growth phase, while ethanol is produced primarily as a non-growth-related product (Klasson 1992). The production of acetate is favored at higher pH (5–7), whereas the production of ethanol is favored at lower values (pH 4–4.5). *C. ljungdahlii* grows on syngas at pH 4–7 producing ethanol and acetate (Tanner et al. 1993). It has been used in full-scale production of ethanol from syngas in a commercial process involving gasification, fermentation, and distillation.

A few years after *C. ljungdahlii* was described, *Clostridium autoethanogenum*, another acetogen able to produce ethanol from CO, was isolated from rabbit feces (Abrini et al. 1994). *C. autoethanogenum* was isolated from rabbit feces in 1994 and has a reported ideal growth temperature of 37 °C (Abrini et al. 1994). *C. autoethanogenum* is a strictly anaerobic, Gram-positive, spore-forming, rodlike, motile bacterium which metabolizes CO to form ethanol, acetate, and CO$_2$ as end products. It is also capable of using CO$_2$ and H$_2$, as well as organic compounds such as pyruvate, xylose, arabinose, fructose, rhamnose, and L-glutamate as substrates (Abrini 1994). Minimal research was done on *C. autoethanogenum* as a gas-fermenting organism until the past 5 years when it has undergone research for the production of ethanol with synthesis gas or pure carbon monoxide as feedstock (Cotter et al. 2009a, b; Guo et al. 2010; Abubackar et al. 2012). The conversion efficiency of *C. autoethanogenum* from syngas to alcohol and acetate, however, was much lower than that of *C. ljungdahlii* (Cotter et al. 2009a). Only low-level ethanol production of 0.32 g L$^{-1}$ (Cotter et al. 2009b), 0.28 g L$^{-1}$ (Abrini et al. 1994), and 0.26 g L$^{-1}$ (Guo et al. 2010) has been reported for *C. autoethanogenum*, with CO as the sole carbon source. *C. ljungdahlii* and *C. autoethanogenum* were among the first organisms identified that convert CO, CO$_2$, and H$_2$ (syngas) to ethanol and acetic acid (Abrini 1994; Vega 1990).

The only published research showing growth and ethanol production from actual biomass producer gas has been done with the acetogen *Clostridium carboxidivorans*

P7 (Datar et al. 2004; Liou et al. 2005). *C. carboxidivorans* P7 (originally named bacterium P7) was isolated from an agricultural settling lagoon and was extensively studied because of its ability to produce six times more ethanol than acetate (Rajagopalan et al. 2002). It was found to produce acetate, ethanol, butyrate, and butanol from CO to $H_2$ (Liou et al. 2005). It is motile, Gram-positive, and spore-forming and forms acetate, ethanol, butyrate, and butanol as end products. The optimum pH range for this strain is 5.0–7.0, and the optimum temperature range is 37–40 °C. Currently, the most attractive syngas fermentation process is using *C. carboxidivorans* P7 to produce butanol (see Sect. 4.2.1.3), which is of higher energy content, lower vapor pressure, and less corrosive than ethanol.

*Clostridium ragsdalei* or strain P11 was isolated from duck pond sediment by researchers from The University of Oklahoma and Oklahoma State University and is described in a patent (Huhnke et al. 2010). *C. ragsdalei* has been explored for the production of ethanol from syngas (Saxena and Tanner 2011; Kundiyana et al. 2011a, b), with growth temperatures of 32–37 °C (Kundiyana et al. 2011b) and a batch fermentation reported ethanol concentration of 1.99 g $L^{-1}$ (Saxena et al. 2012). In a 100-L STR, an ethanol concentration of 25.26 g $L^{-1}$ was achieved over fermentation duration of 59 days (Kundiyana et al. 2010).

*Acetobacterium bacchi* was isolated from livestock-impacted soil in 2010 and has been recently investigated for the production of ethanol from syngas, with a reported ideal growth temperature of 37 °C (Allen et al. 2010). This was notably carried out at an initial pH between 7.7 and 8.0, with *A. bacchi* moderately alkaliphilic (Liu et al. 2012). *A. bacchi* strain CP15 achieved a maximum reported productivity of 1.7 g $L^{-1}$ with 76 % ethanol yield from utilized CO with pure coal-derived syngas (40 % CO, 30 % $CO_2$, and 30 % $H_2$). Using biomass syngas (20 % CO, 15 % $CO_2$, 5 % $H_2$, and 60 % $N_2$), ethanol yield from utilized CO has been reported at 65 % (Liu et al. 2012).

Thermophilic Carboxydotrophs that Produce Ethanol

*Moorella* sp. HUC22-1 is a thermophilic microorganism capable of producing ethanol from synthesis gas (Sakai et al. 2004); however, the ethanol produced by this organism is low. Even when the pH and the cell recycle were lowered, the ethanol production was improved only 15-fold (1–15 mM) and an ethanol:acetate molar ratio of 1:45 (Sakai et al. 2005). An acetaldehyde dehydrogenase (Aldh) and three alcohol dehydrogenases (AdhA, AdhB, and AdhC) have been described from this organism (Inokuma et al. 2007; Sakai et al. 2004). Aldh was shown to catalyze the thioester cleavage of acetyl-CoA, as well as the thioester condensation from CoASH and acetaldehyde. It also was shown to have activity toward both NADP (H) and NAD(H), but activity toward NAD(H) was determined to be eightfold higher. AdhA was observed to catalyze the NADP(H)-dependent reduction of acetaldehyde as well as the oxidation of ethanol. This enzyme can also use NAD(H) as a cofactor but with decreased activity. AdhB was active only when NADP(H) was used as a cofactor. AdhC showed no activity with any of the cofactors used.

Both AdhA and AdhB were active toward reduction of a variety of aldehydes. Surprisingly, the highest activities were toward *n*-butylaldehyde and isobutylaldehyde, even though this organism has not been shown to produce butanol. Finally, the study reported higher Aldh gene expression when cells were grown on $H_2/CO_2$, but lower adhABC expression in cells grown on $H_2/CO_2$ than cells grown on fructose (Inokuma et al. 2007; Sakai et al. 2004).

### 4.2.1.3  Carboxydotrophs that Produce Butanol from CO or Syngas

Mesophilic Carboxydotrophs that Produce Butanol

The industrial acetone–butanol–ethanol (ABE) fermentation was the second largest fermentation process in history behind ethanol fermentation, using sugar or starch utilizing solventogenic clostridia as *Clostridium acetobutylicum*, *C. beijerinckii*, *C. saccharobuylicum*, or *C. saccharoperbutylacetonicum* (Köpke et al. 2011; Dürre 2007). *C. acetobutylicum* has been the model organism for research in ABE fermentation from sugars.

*Butyribacterium methylotrophicum* is a catabolically versatile spore-forming anaerobe (Zeikus et al. 1980; Lynd et al. 1982; Kerby and Zeikus 1987; Worden et al. 1989; and Grethlein et al. 1991). It is an anaerobe capable of fermenting single carbon substrates (CO, $CO_2/H_2$, $CH_4$, and formate) and multi-carbon substrates (i.e., glucose, lactate, pyruvate, sucrose, and glycerol) into varying proportions of acetic acid, butyric acid, ethanol, or butanol (Zeikus et al. 1980). It also possesses the advantageous ability to produce butanol from synthesis gas (Grethlein et al. 1990; Lynd et al. 1982; Zeikus et al. 1980). It is one of the most versatile CO-utilizing bacteria (Grethlein et al. 1991). Other fermentation products are ethanol, acetate, and butyrate. Notably, growth on high CO (more than 101 kPa, 1 atm) concentrations and the production of butyrate and butanol from CO were dependent on the selection of a CO-adapted strain. The first attempts at investigating this strain for butanol production from CO yielded concentrations of 1.4 g $L^{-1}$ (Worden et al. 1991). After some changes in fermentation setup, such as operation at pH 5.5 and continuous cell cycle, butanol production from CO was improved by more than 200 % (Grethlein et al. 1991). Nevertheless, with classic ABE strains producing butanol at more than 400 g $L^{-1}$ (Lee et al. 2008), *B. methylotrophicum* is not yet a contender for commercial biobutanol production.

Another interesting, but less studied, strain for butanol production is *C. carboxidivorans* P7. Being able to produce up to four times more ethanol than butanol from CO or producer gas, this strain has mostly been studied for its ethanol production capabilities (Datar et al. 2004; Liou et al. 2005; Rajagopalan et al. 2002). *C. carboxidivorans* or strain "P7" was isolated in 2005 from a lagoon settlement and can grow autotrophically with $H_2$ and $CO_2$, or CO as substrate, or heterotrophically with simple sugars, having an optimal growth temperature of 37–40 °C (Liou et al. 2005). Products include acetate, ethanol, butanol, and butyrate. Draft genome sequences are available for *C. carboxidivorans* (Paul et al. 2010; Hemme et al. 2010), which contain the genes of the reductive acetyl-CoA pathway as well as enzymes for the conversion of acetyl-CoA into butanol and butyrate. Bruant et al. (2010) found that the strain contains a plasmid and a butanol pathway similar to that of *C. acetobutylicum* (Bruant et al. 2010). The two-carbon acetyl segment of acetyl-CoA is converted to the four carbon butyryl-CoA through thiolase, 3-hydroxybutyryl-CoA dehydrogenase, crotonase, and butyryl-CoA dehydrogenase as in ABE fermentation organism *C. acetobutylicum*, while CoA transferase genes for acetone production are absent in *C. carboxidivorans* (Bennett 1995). Butyryl-CoA is then converted to butyrate and butanol in a similar manner to acetate and ethanol from acetyl-CoA; enzymes such as alcohol dehydrogenase are often unspecific and act to produce both butanol and ethanol. *C. carboxidivorans* has been explored for the production of ethanol (Hurst et al. 2010; Ukpong et al. 2012) as well as the production of butanol (Bruant et al. 2010).

*Clostridium scatologenes* was isolated from soil in 1925, but the type strain was not originally described as acetogenic (Küsel et al. 2000). Although acetate is the primary end product, butyrate is also produced from sugars (Küsel et al. 2000). *C. scatologenes* has an optimal growth temperature of 37–40 °C (Liou et al. 2005).

*Clostridium drakei* was isolated from an acidic coal mine pond (Liou et al. 2005) and is similar to *C. carboxidivorans* and *C. scatologenes*. Originally classified as strain *C. scatologenes* SL1 (Liou et al. 2005; Küsel et al. 2000), *C. drakei* has an optimal growth temperature of 37 °C.

Thermophilic Carboxydotrophs that Produce Butanol

Nguyen et al. (2013) isolated 250 thermophilic bacteria from manure composts. Of these, 4 isolates (T1–16, T2–22, T3–14, and T7–10) were able to use CO as the sole source of carbon and energy. To assess the biochemical basis for their ability to produce butanol from CO, CODH and butanol dehydrogenase activities were assessed for each of the isolates. All isolates showed evidence of CODH and BDH enzyme activities, with the majority exhibiting higher activities compared with the known carboxydotroph, *B. methylotrophicum*. The level of activities for CODH and BDH ranged from 0.163–3.59 and 0.19–2.2 $\mu$mol min$^{-1}$, respectively (Nguyen et al. 2013). Even though several anaerobic carboxydotrophs have been isolated with the ability to convert synthesis gas to biofuels, they are predominantly mesophilic (Table 4.3). So far, very few attempts have been made to isolate thermophilic microorganisms that can produce organic compounds from syngas. Growth temperatures by thermophiles at high temperatures could be advantageous as less cooling of the syngas is required before it is introduce into the bioreactor. Moreover, higher temperatures can lead to higher conversion rates, although higher temperatures do have a negative impact on the solubility of CO and $H_2$ (Henstra et al. 2007b).

#### 4.2.1.4 Carboxydotrophs that Produce Hydrogen from CO or Syngas

Carboxydotrophs that Produce Hydrogen

Biological hydrogen production is environmentally friendly and requires less energy input compared to chemical processes (Ismail et al. 2008). *Rhodospirillum rubrum*, *Rhodopseudomonas palustris* P4, *Citrobacter* sp., and *Rubrivivax gelatinosus* CBS are photoautotrophic and chemoheterotrophic microorganisms capable of performing the gas–water shift (GWS) reaction (Klasson et al. 1992b; Jung et al. 1999; Maness and Weaver 2002; Najafpour et al. 2003, 2004; Markov and Weaver 2008; Ismail et al. 2008). Two enzymes that mainly contribute to the GWS reaction in these organisms are CODH and hydrogenase. The former catalyzes the oxidation of CO, and hydrogenase mediates the reduction of protons to $H_2$ (Maness et al. 2005; Najafpour et al. 2004). Biological GWS is thermodynamically favorable at room temperature ($CO + H_2O \rightarrow H_2 + CO_2$, $\Delta G = -20$ kJ mol$^{-1}$) and atmospheric pressure, compared to the chemical catalysis where a two-stage process is required as well as high temperature (>200 °C) (Benemann 1999). Therefore, minimum energy requirements and low process cost are expected (Ismail et al. 2008).

*Rhodospirillum rubrum* grows quickly and reaches high cell concentrations that uptake CO more rapidly than other similar organisms capable of performing the GWS reaction (Klasson et al. 1992; Najafpour et al. 2003, 2004). It also tolerates small amounts of $O_2$ and sulfur often present in syngas (Klasson et al. 1992). As a result, this strain is the favorite organism for studies investigating biohydrogen production from syngas. *R. rubrum* requires a light source for growth; however, $H_2$ production is independent of light intensity (Najafpour et al. 2004). An organic carbon source is needed for this organism to efficiently consume CO. The highest CO consumption (90–97 %) has been determined to occur when *R. rubrum* is provided with acetate as the organic carbon substrate (1–2 g L$^{-1}$), resulting in a 98 % hydrogen production yield (Najafpour et al. 2004; Najafpour and Younesi 2007). Investigations involving CSTR of *R. rubrum* and with continuous CO flow resulted in hydrogen yields and CO conversions of 87 and 95 % theoretical values, respectively (Younesi et al. 2008). This type of bioreactor was stable for continuous operation for 27 days (Ismail et al. 2008). Most hydrogen production from syngas research has been done using artificial syngas. A mixture of pure gases is normally present in synthesis gas at a fixed concentration (e.g., 56.0 % $N_2$, 17.2 % CO, 16.3 % $CO_2$, and 8.8 % $H_2$).

*Rubrivivax gelatinosus* CBS is another promising strain for its use in syngas-to-hydrogen conversion. In the presence of CO, this organism carries out the WGS reaction in both light and dark conditions (Maness and Weaver 2002). These cells are capable of converting 100 % of CO in the gas phase to $H_2$ in the dark (Markov and Weaver 2008). Its tolerance for oxygen (Maness and Weaver 2002) and its capacity to use CO as the sole carbon and energy source (Markov and Weaver 2008) make it an attractive biocatalyst.

Thermophilic Carboxydotrophs that Produce Hydrogen

A rapidly increasing group of carboxydotrophic hydrogenogenic prokaryotes is formed by strict anaerobic thermophiles (Table 4.3). Conversion of CO to $H_2$ at elevated temperatures has been observed in freshwater as well as marine environments with temperatures ranging from 55 to 85 °C and pH between 5.5 and 8.5 (Bonch Osmolovskaya et al. 1999; Henstra et al. 2007b). All isolated species are capable of chemolithotrophic growth on CO. So far, there exists no evidence of growth inhibition by high levels of CO. Some isolates also grow by fermentation or anaerobic respiration. Because of simultaneous $H_2$ production and acceptor reduction, it is unknown whether CO is a direct electron donor in anaerobic respiration or $H_2$ acts as an intermediate. *C. hydrogenoformans* was originally described as obligate carboxydotroph (Svetlichny et al. 1991b). Later it was found capable to ferment pyruvate to acetate (Svetlichnyi et al. 1994), as well as more recently discovered that it could respire anaerobically CO as electron donor with different electron acceptors, e.g. iron, nitrate, and quinones (Henstra and Stams 2004). *Caldanaerobacterium subterran*eus subsp. *pacificus* (previously *Carboxydobrachium pacificum*) is the only marine carboxydotrophic bacterial species described so far (Fardeau et al. 2004; Sokolova et al. 2001). It was isolated from a submarine hydrothermal vent. Besides CO, it also grows organotrophically on several mono- and disaccharides, cellulose, and starch. *T. carboxydivorans* is the only known species with a Gram-negative cell wall in this group of thermophilic carboxydotrophs (Sokolova et al. 2004b).

## 4.2.2   Biochemistry of Syngas Fermentation

The reductive Acetyl-CoA or the Wood–Ljungdahl pathway (Fig. 4.15) is exclusively used by anaerobic microorganisms for conservation of energy, and for assimilation $CO_2$ into cell carbon (Henstra et al. 2007b; Drake et al. 2008). It was first described in the acetogen *M. thermoacetica* (formerly known as *C. thermoaceticum*), a heterotroph capable of producing 3 mol of acetate from 1 mol of glucose. In addition to the reductive acetyl-CoA pathway, four other biological pathways are known for complete autotrophic $CO_2$ fixation: the Calvin cycle, the reductive tricarboxylic acid (TCA) cycle, the 3-hydroxypropionate/malyl-CoA cycle, and the 3-hydroxypropionate/4-hydroxybutyrate cycle (Thauer 2007). The reductive acetyl-CoA pathway is not cyclic like the Calvin cycle or the reverse TCA cycle in that it is formed by two branches (Fig. 4.15). These branches have been called the Eastern (or carbonyl) and Western (or methyl) branches and were the research focus of Lars Ljungdahl and Harland Wood, respectively (Ragsdale 1997). The Eastern branch produces the methyl group of acetyl-CoA, and the Western produces the carbonyl group. The methyl group is formed from the reduction of $CO_2$ to formate which is converted to formyltetrahydrofolate and is reduced to methyltetrahydrofolate. These steps involve FDH and a series of tetrahydrofolate-dependent enzymes (Drake 1994). Scientists

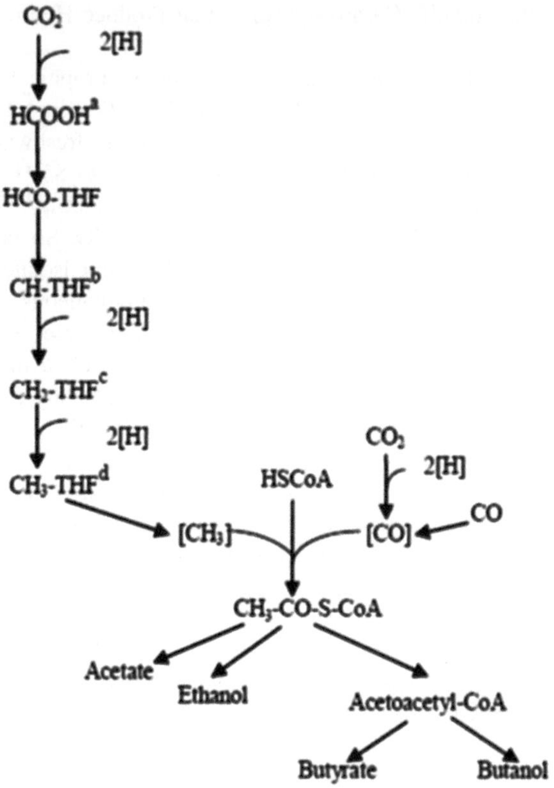

**Fig. 4.15** Schematics of the reductive acetyl-CoA/ Wood–Ljungdahl pathway. [a]Formate (HCOOH), [b]methylene-tetrahydrofolate (CH-THF), [c]methenyl-tetrahydrofolate (CH$_2$-THF), and [d]methyl-tetrahydrofolate (CH$_3$-THF)

have suggested that before an $O_2$ atmosphere appeared, the first autotrophs on Earth are able to use CO or $CO_2$ as their sole source of carbon and thus utilize the acetyl-CoA pathway (Ragsdale and Wood 1991; Pereto et al. 1999; Russell and Martin 2004). The reductive acetyl-CoA pathway is biochemically the simplest among the autotrophic pathways and postulated to be the first autotrophic process on earth (Lindahl and Chang 2001; Drake et al. 2006). This ancient pathway is diversely distributed among at least 23 different bacterial genera: *Acetitomaculum, Acetoanaerobium, Acetobacterium, Acetohalobium, Acetonema, Alkalibaculum, "Bryantella," "Butyribacterium," Caloramator, Clostridium, Eubacterium, Holophaga, Moorella, Natroniella, Natronincola,* Oxobacter, *Ruminococcus, Sporomusa, Syntrophococcus, Tindallia, Thermoacetogenium, Thermoanaerobacter,* and *Treponema* (Drake et al. 2008).

CODH is the central enzyme in this pathway (Wood et al. 1986), and it has also been characterized in a diverse group of organisms. In acetogens, this enzyme is responsible for reducing $CO_2$ to CO yielding the carbonyl group of acetyl-CoA, and it also catalyzes the final step in synthesizing acetyl-CoA from $CH_3$, CO, and S-CoA. Acetogens convert acetyl-CoA to acetate gaining an ATP. Some of these organisms can also reduce acetyl-CoA to acetaldehyde and ethanol using electron donors such as NAD(H) and NADP(H). This results in a net ATP consumption

(Klasson et al. 1992). In chemolithoautotrophs, this enzyme also enables the utilization of CO as the sole carbon and electron source by catalyzing the oxidation of CO to $CO_2$. In these organisms, energy is conserved through an ETC. In hydrogenogenic carboxydotrophs, the CODH reaction is the same as chemolithoautotrophs. However, the electrons are transferred to a membrane-associated hydrogenase that combines the generation of hydrogen with the translocation of protons. This generates the proton gradient needed for ATP formation by ATP synthase. In acetoclastic archaea, the CODH works in a reverse manner in which acetyl-CoA is formed from acetate. Acetyl-CoA is cleaved, the methyl group is reduced to $CH_4$, and CO is oxidized to $CO_2$. Energy in this system is also generated through an ETC. In methanogens able to grow on $CO_2/H_2$ or CO, CODH drives the formation of acetyl-CoA from methyltetrahydrosarcinapterin and CO as well as the oxidation of CO to $CO_2$. In *M. barkeri,* a hydrogenase couples generation of hydrogen with the translocation of protons much like the process that occurs in hydrogenogenic carboxydotrophs. In sulfate-reducing bacteria, CODH functions very similar to the reverse acetyl-CoA pathway in methanogens. More recently, there has been evidence of a CODH enzyme in hyperthermophilic archaea able to grow on CO. These CODHs are very similar to counterparts in methanogens, such as *Methanosarcina acetivorans* C2A and *Methanosarcina mazei* Go¨1 (Lee et al. 2008). There has even been found a hyperthermophilic bacterium, *C. hydrogenoformans*, containing five different forms of this enzyme (Wu et al. 2005a).

Products from syngas fermentation are limited to these reactions and can conserve enough energy for microbes' metabolism from syngas. The Wood–Ljungdahl pathway (Fig. 4.16) is the main mechanism for the production of acetate, ethanol, butyrate, and butanol from syngas fermentation by acetogens, sulfate-reducing bacteria, and archaea under strictly anaerobic condition. CO enters the pathway through two routes. One molecule directly enters the Western branch as CO, while another molecule of CO is oxidized to $CO_2$ by a monofunctional CODH in the BWGSR, with the resulting energy of this reaction being captured as reduced ferredoxin (Drake et al. 1980; Shanmugasundaram and Wood 1992). Since CO is abundant in syngas, the bifunctional CODH complex is not manifestly required. Some of the resulting $CO_2$ then enters the Eastern branch of the reductive acetyl-CoA pathway. This depends if CO serves as both carbon and energy source, or if an additional energy source such as hydrogen is present which can be utilized in a hydrogenase reaction (Fig. 4.16). The electron production is thermodynamically more favorable from CO than from $H_2$ (Hu et al. 2011), and hydrogenases are reversibly inhibited by CO (Bennett et al. 2000; Greco et al. 2007; Matsumoto et al. 2011). Thus, at high CO concentrations, no or only little hydrogen uptake will occur, but it will increase, once CO is utilized and the concentration drops. Ethanol and acetate can be produced according to the following reactions: with CO as the sole carbon an energy source, as in Eqs. (4.1) and (4.2); with CO as a carbon source and both CO and $H_2$ as the energy source, according to Eqs. (4.3–4.5); and with $CO_2$ as carbon source and $H_2$ as energy source, as in Eqs. (4.6) and (4.7) (Ljungdahl 1986):

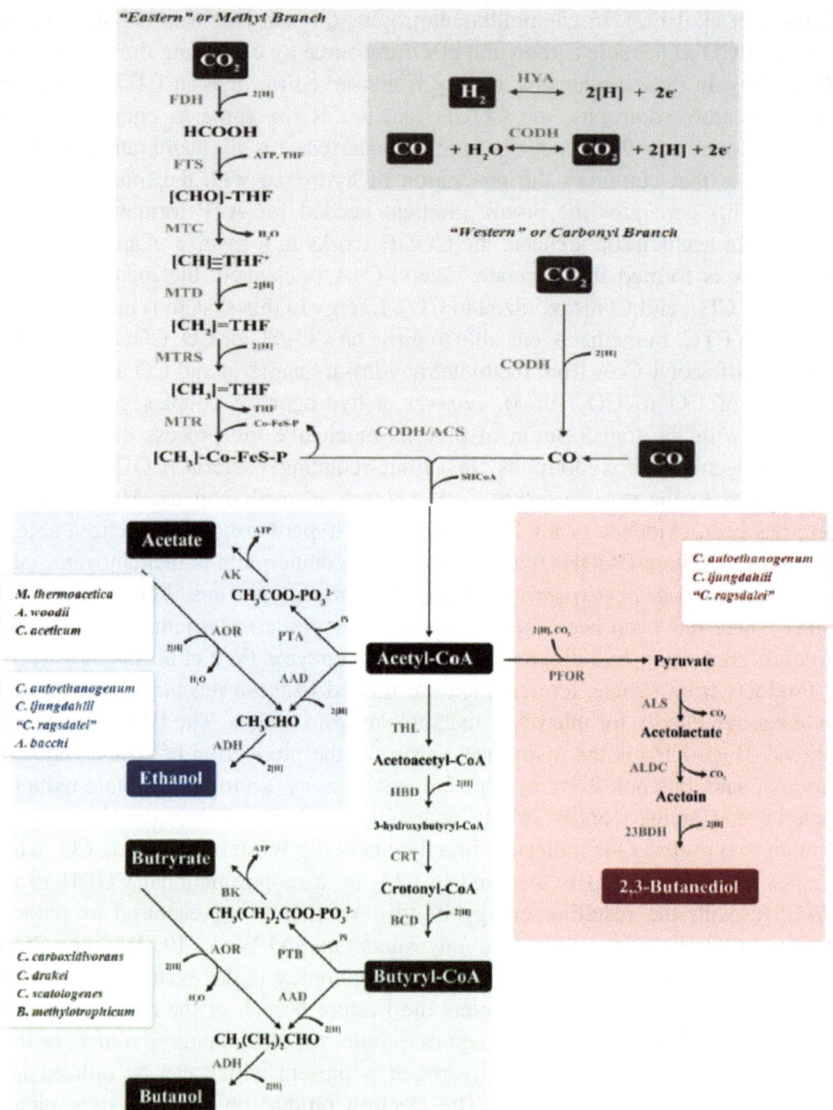

**Fig. 4.16** Wood–Ljungdahl pathway of acetogens and their metabolic end

$$6\,CO + 3\,H_2O \rightarrow CH_3CH_2OH + 4\,CO_2 \quad (\Delta G^{\circ\prime} = -224\,kJ/mol) \qquad (4.1)$$

$$4CO + 2H_2O \rightarrow CH_3COOH + 2CO_2 \quad (\Delta G^{\circ\prime} = -175\,kJ/mol) \qquad (4.2)$$

$$3CO + 3H_2 \rightarrow CH_3CH_2OH + CO_2 \quad \left(\Delta G^{\circ\prime} = -164\,kJ/mol\right) \qquad (4.3)$$

$$2CO + 2H_2 \rightarrow CH_3COOH \quad \left(\Delta G^{\circ\prime} = -135\,kJ/mol\right) \qquad (4.4)$$

$$2CO + 4H_2 \rightarrow CH_3CH_2OH + H_2O \quad \left(\Delta G^{\circ\prime} = -144\,kJ/mol\right) \qquad (4.5)$$

$$2CO_2 + 6H_2 \rightarrow CH_3CH_2OH + 3H_2O \quad \left(\Delta G^{\circ\prime} = -104\,kJ/mol\right) \qquad (4.6)$$

$$2CO_2 + 4H_2 \rightarrow CH_3COOH + 2H_2O \quad \left(\Delta G^{\circ\prime} = -95\,kJ/mol\right) \qquad (4.7)$$

In the Eastern (or methyl) branch, FDH reduces $CO_2$ to formate (Fig. 4.16), which is then attached to tetrahydrofolate (THF) by 10-formyl-THF synthetase (Ragsdale and Pierce 2008). This undergoes several reductive steps catalyzed by enzymes including methylene-THF cyclohydrolase (MTC), methylene-THF dehydrogenase (MTD), and methylene-THF reductase (MTRS). Methyltransferase (MTR) then transfers the methyl group from methyl-THF to a corrinoid-FeS protein (Bruant et al. 2010), and then, this methyl group is provided as the methyl group of Acetyl-CoA. The genes encoding the enzymes that operate in the Eastern branch are ubiquitous and in *M. thermoacetica* are dispersed throughout the genome (Ragsdale and Pierce 2008), while found in a large single cluster in *C. ljungdahlii* (Köpke et al. 2010), *C. autoethanogenum* (Köpke et al. 2011), *C. ragsdalei* (Köpke et al. 2011), *C. carboxidivorans* (Bruant et al. 2010), and *A. woodii* (Poehlein et al. 2012) with the exception of the genes for FDH. The Western (or carbonyl) branch (Fig. 4.16) is unique to anaerobic microorganisms (Ragsdale 1997). CO can either be used directly, or generated from $CO_2$, and serves as the carbonyl group for acetyl-CoA synthesis. The unique multi-subunit bifunctional metaloenzyme CODH/ACS is a characteristic and name-giving feature of the pathway (Doukov et al. 2002). This key enzyme is capable of reducing $CO_2$ to CO in the Western branch, and accepting the methyl group of the corrinoid-Fe/S-protein of the Eastern branch and condensing both the methyl and the carbonyl moiety with a CoA group to produce a molecule of coenzyme-A. Acetyl-CoA-synthase/CO-dehydrogenase complex (ACS/CODH) is responsible for forming acetyl-CoA by joining a carbonyl and a methyl group (Lindahl et al. 2002; Ragsdale 2004; Henstra et al. 2007b). Bacteria and archaea have slightly different acetyl-CoA pathway. For bacteria, formate is reduced from $CO_2$ and then forms formyl compounds bounded to the pterin THF with the utilization of ATP (Henstra et al. 2007b). For Archaea, a methanofuran-bound formyl reduced from $CO_2$ is converted to tetrahydromethanopterin. Because the formation of acetyl-CoA from $H_2$ and $CO_2$ requires energy, acetate is transferred from acetyl-CoA to recover the lost energy from the formation of acetyl-CoA. Ethanol is produced from the further reduction of acetate. Two acetyl-CoA molecules yield an acetoacetyl-CoA which further produces butanol and butyrate (Henstra et al. 2007b; Ragsdale and Pierce 2008; Fischer et al. 2008).

Hydrogen production from syngas is metabolized by hydrogenogenic carboxydotrophs (Fig. 4.17). The acetyl-CoA pathway with the energy-balanced

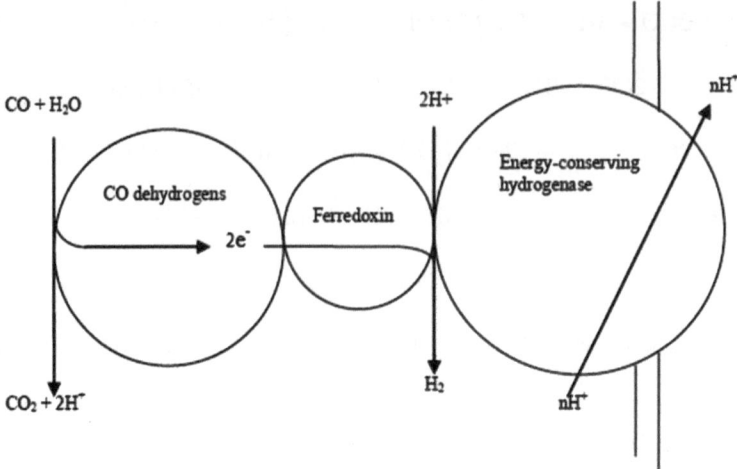

CO + H₂O

2H+

nH⁻

CO dehydrogens

Ferredoxin

Energy-conserving hydrogenase

2e⁻

CO₂ + 2H⁻

H₂

nH⁺

nH⁺

**Fig. 4.17** Schematics of membrane-bound CO-oxidizing, hydrogen-evolving enzyme complex of carboxydotrophic hydrogen (Henstra et al. 2007b)

conservation governs this bacterial metabolism (Henstra et al. 2007b). A monofunctional CODH releases electrons by oxidizing CO to $CO_2$. Energy-converting hydrogenase (ECH) receives electrons and reduces protons to yield hydrogen (Hedderich 2004; Singer et al. 2006). ECH, which forms $H_2$ and translates protons or sodium ion, produces ATP from an ATP synthase driven by a chemiosmotic ion gradient (Hedderich 2004; Maness et al. 2005).

## *4.2.3  Products of Syngas Fermentation*

### 4.2.3.1  Acetate from Syngas

Acetic acid is an important industrial feedstock which has been traditionally produced from petrochemical feedstocks through methanol carbonylation or acetaldehyde oxidation (Wagner 2002). Used as a starting material for vinyl acetate and acetic anhydride synthesis (Wagner 2002), demand for acetic acid has grown over the past decade and is expected to reach 12.15 million tons by 2017 (Global Industry Analysts Inc. 2012). Three organisms have been used for acetate or calcium–magnesium–acetate (CMA) production: *C. aceticum, M. thermoacetica,* and *A. woodii* (Parekh 1991; Drake 1994; Sim et al. 2007; Demler and Weuster-Botz 2011). While acetic acid has no potential as fuel, some organisms such as oleaginous yeasts are able to convert acetic acid and other volatile fatty acids (VFAs) into lipids, which could be used as biodiesel (Fei et al. 2011; Jin et al. 2012). A process scheme has been proposed combining such a fermentation process with a $CO_2$ fermentation process (Stephanopoulos 2011). An acetogen converts $CO_2$ to acetate, which is then used by oleaginous yeast to produce lipids with $CO_2$ as a by-product which could then be fed back to the acetogen. The reducing power can either come from $H_2$ or electricity.

### 4.2.3.2 Ethanol from Syngas

Ethanol is by far the biofuel produced in the greatest quantities worldwide (Demirbas and Balat 2006). Brazil is the number one ethanol producer in the world with 41 % of the total production, followed very closely by the United States (Herrera 2006). Most of this ethanol is produced by microbial fermentation of sugars from either sugarcane or corn starch. To make ethanol a commercial fuel contender, the feedstock has to be switched to lignocellulosic biomass. Hydrolysis–fermentation technology has proven to be expensive and labor-intensive. Up to 40 % of the carbon present in the biomass is lost in the form of lignin, and most microorganisms used in this process are unable to utilize 5-carbon sugars in the hydrolysates. Moreover, ethanol produced this way has not been able to compete with fossil fuel derivatives such as gasoline and diesel. Biomass gasification can yield up to 100 % carbon conversion to gas components, and fermentation of syngas to ethanol has been shown to be commercially feasible (Tirado-Acevedo et al. 2010). In 1987, *C. ljungdahlii* was discovered to have an ability to ferment carbon monoxide and hydrogen into ethanol and acetic acid. Since then, there has been significant development in syngas fermentation research, especially in process microbiology with discovery of over dozens of new species and process engineering such as new reactor design for improved mass transfer among others (Lynd 2008). The syngas fermentation into ethanol and other bioproducts is considered to be more attractive due to several inherent merits over the biochemical approach and the Fisher–Tropsch (FT) process (Bredwell et al. 1999; Brown 2003; Heiskanen et al. 2007; Klasson et al. 1990), such as (1) utilization of the whole biomass including lignin irrespective of the biomass quality; (2) elimination of complex pretreatment steps and costly enzymes; (3) higher specificity of the biocatalysts; (4) independence of the $H_2$:CO ratio for bioconversion; (5) aseptic operation of syngas fermentation due to generation of syngas at higher temperatures; (6) bioreactor operation at ambient conditions; and (7) no issue of noble metal poisoning.

Ethanol is the most desirable product in biomass-derived syngas fermentation (Schmidt et al. 2009). Younesi et al. (2005) examined the ethanol and acetate production during syngas fermentation using *C. ljungdahlii*. Under normal growth conditions, the wild-type strain of *C. ljungdahlii* was found to produce mainly acetic acid with an ethanol-to-acetate ratio of 0.05, and ethanol concentrations of 60.1 g $L^{-1}$ (Vega et al. 1990). Several studies focused on improving the ethanol yield in *Clostridia* through the addition of a reducing agent, the supplementation of additional media constituents, pH shifts, the addition of hydrogen, and providing nutrient-limiting conditions. A condition that induces sporulation has also been found to favor solventogenesis (Klasson et al. 1990). The addition of yeast extract

**Table 4.5** Maximum product and cell yields from different studies (Munasinghe and Khanal 2010)

| Microorganism | Ethanol (g L$^{-1}$) | Acetate (g L$^{-1}$) | Yield (g cell g$^{-1}$) | References |
|---|---|---|---|---|
| *Clostridium ljungdahlii* | 48.0 | 3.0 | 0.45 | Klasson et al. (1993) |
| *Clostridium ljungdahlii* | 3.0 | 2.0–3.0 | nd | Klasson et al. (1990) |
| *Clostridium ljungdahlii* | 0.062[a] | 0.094[a] | 1.378[b] | Phillips et al. (1994) |
| Bacterium P7 | 0.15[a] | 0.025[a] | 0.25[b] | Rajagopalan et al. (2002) |
| *Clostridium ljungdahlii* | 0.55 | 1.3 | 0.3 | Younesi et al. (2005) |
| *Clostridium ljungdahlii* | 11.0–12.0 | 28.0 | 1.15 | Najafpour and Younesi (2006) |

[a] mol C in products per mol CO consumed
[b] g mol$^{-1}$ of CO; *nd* not determined

(0.02 %) followed by cellobiose produced a molar ethanol-to-acetate ratio of 1.0, with ethanol concentrations as high as 3.0 g L$^{-1}$ (Klasson et al. 1990). The authors further reported an improved molar ethanol-to-acetate ratio (>1.1) with the supplementation of a reducing agent (Benzyl viologen: 30 ppm). Klasson et al. (1993) reported a drastic improvement in the ethanol yield with *C. ljungdahlii* when the pH was dropped to 4.0–4.5 and a nutrient-limited media was used. The corresponding ethanol concentration was as high as 20 g L$^{-1}$ with an acetate concentration of only 2–3 g L$^{-1}$ in a complete-mix reactor using coal-derived syngas (H$_2$: 25–35 %, CO: 40–65 %, CO$_2$: 1–20 %, and CH$_4$: 0–7 %). The authors reported a maximum ethanol concentration of 48 g L$^{-1}$ using this microbe during 560 h of operation. The corresponding acetate concentration was around 3 g L$^{-1}$. Maximum product and cell yields obtained from different studies are summarized in Table 4.5.

### 4.2.3.3 Butanol from Syngas

Butanol, like ethanol, can be produced from fermentable sugars, synthesis gas, and glycerol. Butanol has a number of notable qualities that make it a suitable alternative fuel. Its energy content is 30 % more than ethanol (Qureshi and Ezeji 2008). It can be mixed with gasoline in any proportion or be used as the sole fuel component (100 % butanol) in unmodified car engines (Ramey 2007). It carries less water and, therefore, it can be transported through existing gasoline pipelines (Dürre 2007). Reports of biological butanol formation date back to Louis Pasteur. He reported an alcohol product from a clostridial culture (Dürre 2007). The ABE fermentation was essential during World War I. Acetone was needed to prepare munitions, and it was in great shortage at the time. Production of acetone by fermentation meant a constant supply of acetone to Britain and its allies (Dürre 2007).

Even though the efficiency of butanol production from syngas is very low, syngas fermentation into butanol has several potential merits because the whole biomass, including lignin, can be utilized without the requirement for complex pretreatment and enzyme hydrolysis. Thus, it alleviates some of the problems associated with the utilization of lignocellulosic biomass. Hence, the development of syngas-based processes for butanol production can substantially improve the economics and viability of biobased butanol production if the fermentation efficiency can be enhanced through metabolic engineering and process optimization.

One of the most critical problems in ABE fermentation is solvent toxicity. For instance, clostridial cellular metabolism ceases in the presence of 20 g $L^{-1}$ or more solvents (Woods 1995). This limits the concentration of carbon substrate that can be used for fermentation resulting in low final solvent concentration and productivity. The lipophilic solvent butanol is more toxic than others as it disrupts the phospholipid components of the cell membrane causing an increase in membrane fluidity (Bowles and Ellefson 1985). Moreira et al. (1981) had attempted to elucidate the mechanism of butanol toxicity in C. acetobutylicum. They found that 0.1–0.15 M butanol caused 50 % inhibition of both cell growth and sugar uptake rate by negatively affecting the ATPase activity. Increased membrane fluidity causes destabilization of the membrane and disruption of membrane-associated functions such as various transport processes, glucose uptake, and membrane-bound ATPase activity (Bowles and Ellefson 1985). Also, butanol is the only solvent produced to the level that becomes toxic to the cells during the fermentation of clostridia (Jones and Woods 1986). It has been known that the addition of 7–13 g $L^{-1}$ of butanol to culture medium results in a 50 % inhibition of growth, while the addition of acetone and ethanol up to 40 g $L^{-1}$ reduced growth by 50 % (Jones and Woods 1986).

To develop economical and sustainable processes for biobased butanol production, the metabolic pathways of native producers, such as clostridia, need to be well characterized and optimally redesigned. Constructing synthetic pathways by using genes from various organisms and newly synthesized genes can also facilitate the redesign of metabolic networks to allow efficient production of butanol (Jang et al. 2012). Systems-level approaches, including genome sequencing and engineering, the construction of genome-scale metabolic networks of clostridia, omics studies, transcription machinery engineering, and construction of synthetic pathways, have been used to develop more efficient butanol producers (Huang et al. 2010; Jang et al. 2012). Through these studies, the butanol-producing metabolic pathway originating from clostridia has been further characterized and alternative pathways for butanol formation have been identified. However, much improvement is still needed to make an economically competitive process for biobased butanol production. Based on the genome sequence, a genome-scale metabolic model has been successfully reconstructed. Transcriptome and proteome profiling studies performed under various genotypes and environmental conditions allowed us to decipher some of the underlying mechanisms of gene expression and protein abundance, including regulatory circuits. Although not yet as efficient as that in *E. coli* and other well-studied microorganisms, metabolic engineering tools for gene knockout and amplification in clostridia are available. Integration of all of these strategies for systems-level metabolic engineering of clostridia will allow the development of a superior strain capable of efficiently producing butanol from renewable biomass (Jang et al. 2012).

### 4.2.3.4  Hydrogen from Syngas

While ethanol and butanol have much potential as an alternative fuel source, hydrogen has the potential to be even more efficient as a fuel source. Hydrogen introduces no new carbon to the atmosphere and produces no harmful by-products. Hydrogen holds the promise as a dream fuel of the future with many social, economic, and environmental benefits to its credit. Molecular $H_2$ has the highest energy content per unit weight among the known gaseous fuels ($143$ GJ ton$^{-1}$) (Boyles 1984) and is the only carbon-free fuel which ultimately oxidizes to water as a combustion product. Thus, hydrogen is considered to be a clean fuel since water is the only by-product when it is burned. Burning hydrogen not only has the potential to meet a wide variety of end-use applications but also does not contribute to greenhouse emission, acid rain, or ozone depletion (Kotay and Das 2008). The use of hydrogen will contribute to significant reduction of these energy-linked environmental impacts. Hydrogen can be used either as the fuel for direct combustion in an internal combustion engine or as the fuel for a fuel cell. The largest users of $H_2$, however, are the fertilizer and petroleum industries with, respectively, 50 and 37 % (Elam et al. 2003; Kotay and Das 2008). Sales of hydrogen have increased by 6 % annually in the last 5 years, which is closely related to the increased use of hydrogen in refineries as a result of stricter standards for fuel quality (Elam et al. 2003). Hydrogen will become world's "clean energy choice," joining electricity as a primary energy carrier and providing the foundation for a globally sustainable energy system (Beneman 1996; Demirbas 2007). It has a wide variety of applications, including fuel for automobiles, distributed and central electricity and thermal energy generation (Fig. 4.18).

The main hindrances to hydrogen use as a fuel are the difficulties in transportation and storage, as hydrogen is highly volatile. Although hydrogen is the most abundant element in the universe, it must be produced from other hydrogen-containing compounds such as fossil fuels, biomass, or water. Each method of production requires a source of energy, i.e. thermal (heat), electrolytic (electricity), or photolytic (light) energy (Kotay and Das 2008). Natural gas and coal SMR are the least expensive known technologies for $H_2$ production. This process results in a gas mixture of mainly CO and $H_2$. Subsequently, CO is chemically converted to $CO_2$ by the WGS reaction producing additional $H_2$. Syngas may also be converted to hydrogen via the WGS reaction. At a production cost of up to $50/GJH$_2$, these processes do not make hydrogen a viable replacement for fossil fuels at present (Ismail et al. 2008). While hydrogen seems to be the ideal fuel for the future, more

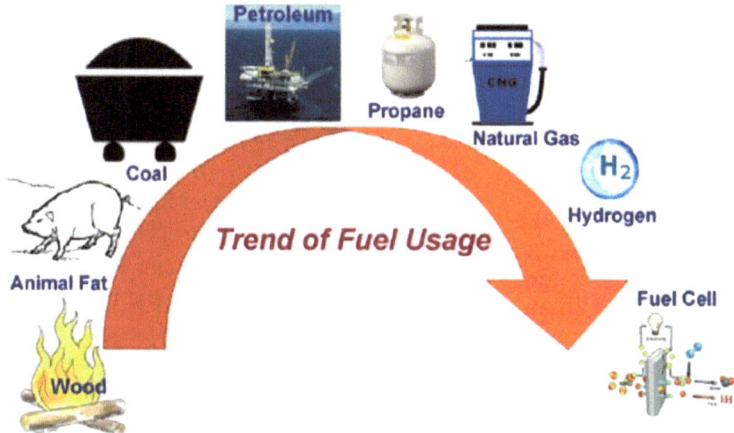

**Fig. 4.18** Chronology of fuel usage by mankind (Kotay and Das 2008)

improvements are necessary to make hydrogen more portable and more easily stored. Future research will likely address these issues.

Microbial $H_2$ production is an attractive process for supplying a significant share of the $H_2$ required for the near future. Several hydrogenogenic carboxydortrophs belonging to the genera Citrobacter, Peptostreptococcus, *Rubrivivax, Rhodopseudomonas, R. rubrum, Moorella, Calderihabitans, Carboxydothermus, Carboxydibrachium, Carboxydocella, Desulfotomaculum, Thermincola, Thermococcus, Thermolithobacter*, and *Thermosinus* are reported in Table 4.3.

Research progress for identifying the major technoeconomic bottlenecks of various bioprocesses for commercial production of hydrogen appears promising. However, high hydrogen yield remains to be the ultimate goal and challenge for the biohydrogen research and development. Enhancement in hydrogen yield may be possible by using suitable microbial strain, process modification, efficient bioreactor design, and also genetic and metabolic engineering technique, to redirect metabolic pathway (Nath and Das 2004).

### 4.2.4 Potential Improvements

While the process of converting waste products to useful fuels is at a marketable stage, there is still room for improvement. One process that could potentially be performed more efficiently is the process in which the syngas is cooled after the gasification stage. While a portion of the excess heat can be converted into electricity, some of the heat is also lost to the surroundings. The discovery or genetic engineering of thermophilic bacteria that are capable of converting carbon monoxide to ethanol would eliminate the need to cool the syngas before it can be converted to ethanol. This would greatly reduce the amount of heat that is lost in the process (Henstra et al. 2007b).

### 4.2.4.1 Bioreactor Design

One area of potential improvement is the reactor design and function. Younesi et al. (2008) performed an experiment very similar to the one performed by Najafpour et al. (2004), using various reactor types to determine the optimal conditions for syngas conversion. In order to maximize syngas conversion, the mass transfer rate must be optimized. In the experiment by Younesi et al. (2008), a microsparger was used to minimize the size of the gas bubbles, and the reactor used was a constant stirred tank bioreactor. Both these factors increased the mass transfer rate. In the experiment, *R. rubrum* was placed in a 2-L bioreactor and exposed to syngas for a period of 2 months. The agitation speed and gas flow rates were varied in order to determine their effects on the mass transfer rate. At low agitation speeds, little carbon monoxide was converted due to the low mixing of reactants. However, when agitation speeds were raised above 500 rpm, the nutrients became toxic due to foaming of the mixture. Low gas flow rates did not provide significant amounts of carbon monoxide to be converted to hydrogen. However, when high gas rates were used, some of the carbon monoxide passed through the reactor too quickly and was not converted. For the experiment by Younesi et al. (2008), it was determined that the conversion of syngas to hydrogen occurred best with a gas flow rate of 14 mL min$^{-1}$ and an agitation speed of 500 rpm (Younesi et al. (2008).

Gas–liquid mass transfer is a rate-limiting step in syngas fermentation process (Worden et al. 1991; Klasson et al. 1993). Mass transfer limitations are inevitable at several points of the diffusion process including the transport of gaseous substrate into gas–liquid interface, its transport into culture media (aqueous phase), the transport of the mixed gases into the stagnant liquid layer around the microbes, and the diffusion of the transported gaseous substrate into the microbial cell. The gas–liquid interface mass transfer is the major resistance for gaseous substrate diffusion. Diffusion limitations of a gaseous substrate into the culture media result in low substrate uptake by microbes and thus lead to low productivity (Munasinghe and Khanal 2010). Reactor configuration is closely related to the gas–liquid mass transfer efficiency. Thus, reactor design plays an important role in syngas fermentation. High mass transfer rates, low operation and maintenance costs, and easy scale-up are some of the key parameters for designing an efficient bioreactor system. Similarly, the bioreactor size greatly depends on the rate of mass transfer for sparingly soluble gases (Vega et al. 1990).

### 4.2.4.2 Microbial Catalysts

The isolation of thermophilic carboxydotrophs capable of converting syngas into biofuels and other bioproducts with higher product yields is another important aspect of commercialization of the process. Recently, metabolic engineering and synthetic biology techniques have been applied to gas fermentation organisms. This work strives to improve microbial productivity and robustness and to introduce pathways for the commercial production of increasingly energy dense fuels and more valuable chemicals. Over the past two decades, genetic techniques for clostridia, such as antisense RNA strategies (Desai et al. 1999), reporter gene systems (Tummala et al. 1999; Girbal et al. 2003; Feustel et al. 2004; Cui et al. 2012), inducible promoter-repressor systems (Girbal et al. 2003; Dong et al. 2012), and several double-crossover homologous recombination strategies (Soucaille et al. 2008; Cartman and Minton 2010; Tracy and Papoutsakis 2010; Tripathi et al. 2010; Argyros et al. 2011; Tracy et al. 2011), have been developed. More recently, integration-based techniques such as ClosTron (Heap et al. 2007, 2010; Kuehne et al. 2011), as well as marker-less integration methods (Heap et al. 2012), have been applied to *C. acetobutylicum* and other species such as the cellulolytic *C. thermocellum*. Several review articles have been published recently that give a detailed overview of the developed tools (Green 2011; Lütke-Eversloh and Bahl 2011; Tracy et al. 2012). Until the last 5 years, there was a notable lack of techniques and tools allowing chromosomal manipulation in gas fermentation organisms. Throughout the literature, many microorganisms have been genetically manipulated for enhanced biofuel production. However, synthesis gas has not been the source of energy or growth for organisms in these experiments. This is probably due to the lack of genetic information and tools for syngas utilizing organisms. The work of Nishio and colleagues (Inokuma et al. 2007) is a good start on identifying what enzymes are more active under syngas fermentation and which ones are not, helping to identify good candidate genes for genetic modifications. Metabolic engineering of these microorganisms may further facilitate biomass conversion to biofuels as well as lowering the cost of these processes. The recent availability of genome sequences for gas fermentation organisms combined with these new molecular biology techniques and transformation protocols developed specifically for clostridia has made the direct modification of gas fermentation organisms possible.

Syngas fermentation is always associated with acid production, which lowers the culture pH. Low pH provides an unfavorable environment for solvent production by clostridia. Redirecting the metabolic pathway toward solvent production by blocking acid production might enhance ethanol production. More research is needed in this area. The use of metabolic engineering to integrate new pathways has been reported in three gas fermentation organisms (Schiel-Bengelsdorf and Dürre 2012; Köpke and Liew 2012): *C. ljungdahlii* and *C. autoethanogenum*, for the production of the biofuel butanol (Köpke et al. 2010; Köpke and Liew 2012), and *C. aceticum*, for the autotrophic production of the chemical acetone (Lederle 2010). These efforts will greatly benefit from systems biology approaches and the creation of genome-scale metabolic models.

### 4.2.4.3 Inhibitory Compounds

Biomass-derived syngas often contains additional constituents such as ethylene ($C_2H_4$), ethane ($C_2H_6$), acetylene ($C_2H_2$), tar, ash, char particles, and gases containing sulfur and nitrogen (Bridgwater 1994; Ahmed et al. 2006; Haryanto et al. 2009). These impurities in the syngas affect the efficiency of the fermentation process by potential scaling in pathways, inhibiting the microbial catalysts resulting in low cell growth and product yield. A study reported cell dormancy, hydrogen-uptake shutdowns, and a shift in pathways from acidogenesis to solventogenesis and vice versa, when the syngas was used without conditioning (Datar et al. 2004). Ahmed and Lewis (2007) was able to overcome cell dormancy by introducing a 0.025-ml filter to remove tar, ash, and other particulate matter from the biomass-derived producer gas. Nitrous oxide (NO) was found to be a potential inhibitor of hydrogenase enzyme activity, which reduced the available carbon for product formation (Ahmed and Lewis 2007). The inhibitory effects of NO on syngas fermentation can be eliminated by improving the gasification efficiency or by scavenging it using agents such as, sodium hydroxide, potassium permanganate, or sodium hypochlorite (Brogren et al. 1997; Chu et al. 2001).

By optimizing pre-treatment and gasification in a feedstock-dependent manner, the formation of impurities can be minimized, reducing the need for expensive gas cleanup. However, since impurities generated can influence variables involved in the fermentation process, including pH, osmolarity, and redox potential and can directly inhibit enzymes and contribute to cell toxicity, a gas cleanup step is important to ensure a clean syngas is produced which does not contain components which will negatively interfere with the fermentation process (Daniell et al. 2012). Impurities in the synthesis and subsequent gas cleanup steps utilized will vary depending on the biomass feedstock (Ahmed et al. 2006). Some gas cleaning techniques include tar cracking, wet cleaning, and the use of activated carbon and ZnO (Boerrigter et al. 2002). Tar cracking techniques include catalytic cracking, thermal cracking, plasma cracking, scrubbing with water, and scrubbing with oil (Rabou et al. 2009). Wet gas cleaning is a conventional method where synthesis gas is in contact with fine droplets of water in a counter or co-current flow. Water-soluble substances are dissolved, including nitric oxide and ammonia. However, new developments using catalytic-based hot-gas cleaning appear to be superior in terms of energy efficiency in removing both tar and ammonia from the synthesis gas (Xu et al. 2010), and ZnO and activated carbon filters are a good method for removing $H_2S$ and other inorganic impurities (Boerrigter et al. 2002). Ahmed et al. (2006) found that synthesis gas from switchgrass produced in a fluidized-bed reactor could be cleaned with a cyclone, 10 % acetone scrubbing bath, and 0.025-µm filters. This process sufficiently cleaned the syngas so that the biocatalyst was viable and the product profile was unaffected, compared with results from a synthetic gas stream (Ahmed et al. 2006).

#### 4.2.4.4  Product Recovery

Although distillation is being used as the traditional method of separating ethanol from a mixture of water and other syngas fermentation by-products, high energy costs challenge the continuation of this method. Ultrasonic atomization, vapor recompression, vapor reuse and vacuum distillation, and selective adsorption of water are some of the alternative methods that have been examined in order to reduce the ethanol recovery cost (Sato et al. 2001). Liquid–liquid extraction is a widely used separation technique for acetic acid recovery. A suitable solvent can be used in order to extract a substantially pure acetic acid solution. Fockedey et al. (2008) proposed a novel extraction/re-extraction method using glycerol as one of the solvents to recover ethanol, acetic acid, and other by-products.

## 4.3  Bioremediation of Toxic Compounds with Thermophilic Carboxydotrophs

Bioremediation involves the use of enzymes and microorganisms to breakdown waste products such that their impact on the environment is minimized (Baker and Herson 1994). The goal in bioremediation is to stimulate microorganisms with nutrients and other chemicals that will enable them to destroy the contaminants (Wu et al. 2005b; Gentile et al. 2006). This technology is considered as an effective and eco-friendly alternative to conventional remediation strategies (Kazy et al. 2006). The bioremediation systems in operation today reply on microorganisms native to the contaminated sites, encouraging them to work by supplying them with the optimum levels of nutrients and other chemicals essential for their metabolism (Margesin and Schinner 2001; Tyagi et al. 2011; Cho et al. 2012). Thus, today's bioremediation systems are limited by the capabilities of the native microbes. However, researchers are currently investigating ways to augment contaminated sites with non-native microbes—including genetically engineered microorganisms—specially suited to degrading the contaminants of concern at particular sites (Crowford and Crowford 2005; Megharaj et al. 2011). It is possible that this process, known as bioaugmentation, could expand the range of possibilities for future bioremediation systems (Tyagi et al. 2011). Regardless of whether the microbes are native or newly introduced to the site, an understanding of how they destroy contaminants is critical to understanding bioremediation. The types of microbial processes that will be employed in the cleanup dictate what nutritional supplements the bioremediation system must supply (Boopathy 2000; Crowford and Crowford 2005; Tiquia 2010; Megharaj et al. 2011). Furthermore, the by-products of microbial processes can provide indicators that the bioremediation is successful.

Microorganisms gain energy by catalyzing energy-producing chemical reactions that involve breaking chemical reactions that involve breaking chemical bonds and transferring electrons away from the contaminant. The type of chemical reaction is called an oxidation–reduction reaction: the organic contaminant is oxidized, the technical term for losing electrons; correspondingly, the chemical that gains the electrons is reduced. The contaminant is called the electron donor, while the electron recipient is called the electron acceptor. The energy gained from these electron transfers is then invested, along with some electrons and carbon from the contaminant, to produce more cells. These electron donor and acceptor are essential for cell growth and are called the primary substrates. Many microorganisms use $O_2$ as the electron acceptor. The process of destroying organic compounds with the aid of $O_2$ is called aerobic respiration. In aerobic respiration, microbes use $O_2$ to oxidize part of the carbon in the contaminants to carbon dioxide ($CO_2$), with the rest of the carbon used to produce new cell mass. In the process, the $O_2$ gets reduced, producing water. Thus, the major by-products of aerobic respiration are $CO_2$, $H_2O$, and an increased population of microorganisms. Many microorganisms can exist without oxygen, using a process called anaerobic respiration. In anaerobic respiration, nitrate ($NO_3^-$), sulfate ($SO_4^{2-}$), metals such as iron ($Fe_3^+$) and manganese ($Mn_4^+$), or even $CO_2$ can play the role of oxygen, accepting electrons from the degraded contaminant. Thus, anaerobic respiration uses inorganic chemicals as electron acceptors. In addition to new cell matter, the by-products of anaerobic respiration may include nitrogen gas ($N_2$), hydrogen sulfide ($H_2S$), reduced forms of metals, and methane ($CH_4$), depending on the electron acceptor.

Bioremediation techniques have been used to treat organic compounds in the effluent from chemical plants, to degrade spills of oil and other fossil fuels, to remove sludge from pipes, to degrade organic matter at land fills, and for a wide variety of other applications (Alexander 1999; Cookson 1995). Efforts have been made to utilize bioremediation processes to convert toxic substances to non-hazardous forms. Bioremediation has the potential advantages of low cost and environmental soundness and has therefore received much attention in the scientific and business communities. Most bioremediation processes are performed in aerobic conditions using aerobic microorganisms. However, many potential contamination sites, such as groundwater contamination, that could benefit from a bioremediation clean-up process are anaerobic in character. Hence, the typical treatment scheme for groundwater contamination, for example, involves withdrawing the groundwater, aerating it, exposing it to aerobic microorganisms to detoxify chemical species within the groundwater, and then returning the water to the ground. Employing such procedures can be expensive and may cause the release of hazardous chemical constituents into the ambient air, thereby defeating the two chief advantages of using bioremediation processes.

CO dehydrogenases and CO dehydrogenase-containing anaerobes hold promise for the bioremediation of toxic compounds (Jablonksi and Ferry 1993; Lowe and Zeikus 1993). Known carboxydotrophs that have been bioremediation and degradation of toxic compounds are reported in Table 4.6. Three of these are thermophilic: *Methanosarcina thermophila* (a methanogen), *C. thermoaceticum* (an acetogen), and *Methanobacterium thermoautrophicum* (a methanogen).

**Table 4.6**   Carboxydotrophs capable of degrading toxic compounds

| Toxic compounds | Microorganism | Temperature classification | References |
|---|---|---|---|
| Pentachlorophenol (PCP) | *Acetobacterium woodii* | Mesophile | Bhatnagar et al. (1989) |
| | *Butyribacterium methylotrophicum* | Mesophile | Bhatnagar et al. (1989) |
| | *Eubacterium limosum* | Mesophile | Bhatnagar et al. (1989) |
| | *Methanobacterium ivanovii* | Mesophile | Bhatnagar et al. (1989) |
| | *Methanobacterium formicum* T1N | Mesophile | Bhatnagar et al. (1989) |
| | *Methanosarcina barkeri* | Mesophile | Bhatnagar et al. (1989) |
| Perchloroethylene (PCP) | *Desulfomonile tiedjei* | Mesophile | Cole et al. (1995), DeWeerd et al. (1990) |
| | *Methanosarcina* strain DCM | Mesophile | Fathepure et al. (1988) |
| | *Methanosarcina mazei* S6 | Mesophile | Fathepure et al. (1987) |
| Trichloroethane (TCA) | *Clostridium* sp. | Mesophile | Galli and McCarty (1989) |
| Tetrachloromethane (CCl$_4$) | *Acetobactenium woodii* | Mesophile | Egli et al. (1988) |
| | *Clostridium thermoaceticum* | Thermophile | Egli et al. (1988) |
| | *Desulfobacterium autotrophicum* | Mesophile | Egli et al. (1988) |
| | *Methanobacterium thermoautotrophicum* | Thermophile | Egli et al. (1990) |
| Trinitrotoluene (TNT) | *Clostridium thermoaceticum* | Thermophile | Huang et al. (2000) |
| Trichloroethene (TCE) | *Methanosarcina thermophila* | Thermophile | Jablonski and Ferry (1992, 1993) |
| Trichlorofluoromethane (CFC-11) | *Methanosarcina barkeri* | Mesophile | Krone and Thauer (1992) |

## *4.3.1  Reductive Dehalogenation of Trichloroethylene by* Methanosarcina Thermophila

Trichloroethylene (TCE) is one of the most frequently cited organic groundwater contaminants and is listed as a priority pollutant by the US Environmental Protection Agency (EPA) (Sittig 1985). Methane-producing microorganisms have been implicated in the reductive dehalogenation of multi-halogenated one-carbon compounds and ethylenes (Fathepure et al. 1987; Fathepure and Boyd 1988; Freedman and Gossett 1989; Krone et al. 1989, 1991). Dehalogenation is the replacement of a dehalogen substituent of a molecule with a hydrogen atom. In all known biological examples of this activity, the halogen (e.g., fluorine, chlorine, bromine, iodine, and astatine) is released as a halide ion. This process makes many xenobiotic compounds less toxic and more readily degradable and appears to be the essential primary step in anaerobic degradation of halogenated aromatic compounds. Anaerobic reductive dehalogenation is the only known biodegradation mechanism of certain significant environmental pollutants, such as highly chlorinated biphenyls, hexachlorobenzene, and tetrachloroethylene. It has been suggested that the reductive dehalogenation of one-carbon compounds may be catalyzed by corrinoids (Krone et al. 1989, 1991) which are present at high levels in methane-producing microorganisms (Gorris et al. 1988). It was reported that reduced vitamin $B_{12}$ reductively dechlorinates TCE (Gantzer and Wackett 1991).

Most of the studies of detoxification of halogenated compounds have involved mixed-culture consortia rather than pure cultures. Many chlorinated organic compounds are not biodegraded under anaerobic conditions, often because the chlorine substitutions prevent ring cleavage and subsequent dechlorination. Certain chlorinated aromatic compounds were shown to be dechlorinated in anaerobic habitats such as sediment, flooded soil, and digested sludge. These chemicals include chlorinated benzoates (Suflita et al. 1982, 1983; Horowitz et al. 1983), chlorinated phenols (Ide et al. 1972; Murthy et al. 1979; Boyd et al. 1983; Boyd and Shelton 1984), some of the pesticides such as diuron [3-(3,4-dichlorophenyl)-1,1-dimethylurea] (Attaway et al. 1982), techlofthalam [N-(2,3-dichlorophenyl)-3,4,5,6-tetrachlorophthalmic acid] (Kirkpatrick et al. 1989), chloronitrofen (4-nitrophenyl-2,4,6-trichlorophenyl ether) (Yamada and Suzuki 1983), and 2,4,5-trichlorophenoxyacetic acid (Suflita et al. 1984). In all of these cases, chlorine is removed from the aromatic ring before ring cleavage, which is in contrast to aerobic metabolism of chlorinated compounds. This reaction (a reductive dechlorination) is therefore important, since it has the potential for making some of the highly chlorinated, serious pollutants less persistent and less toxic.

*Methanosarchla thermophila* is an acetotrophic methane-producing microorganism, which synthesizes high levels of a corrinoid-containing carbon monoxide (CO) dehydrogenase enzyme complex (Terlesky et al. 1986). The enzyme complex cleaves the C–C and C–S bonds of acetyl-CoA (Terlesky et al. 1986) and contains a two-subunit CO-oxidizing nickel/iron–sulfur (Ni/Fe–S) component, and a two-subunit corrinoid/iron–sulfur (Co/Fe–S) component containing the corrinoid, factor III (Abbanat and Ferry 1991). Here, we report on the reductive dechlorination of TCE by the CODH enzyme complex from *M. thermophila*. The purified CODH enzyme complex from *M. thermophila* dechlorinated TCE in the presence of CO (Jablonski and Ferry 1992). After 145 min, approximately one-third of the TCE was transformed to *cis*-DCE, *trans*-DCE, vinyl chloride, and ethylene. A proposed mechanism for reductive dechlorination of TCE to the cis-DCE product by the CODH enzyme complex is presented in Fig. 4.19. This mechanism incorporates recent results which show that the CO-reduced Ni/Fe–S component transfers

**Fig. 4.19** Proposed
mechanism of reductive
dechlorination of TCE by the
CO-reduced CO-
dehydrogenase enzyme
complex from
*Methanosarcina thermophila*
(Jablonski and Ferry 1992)

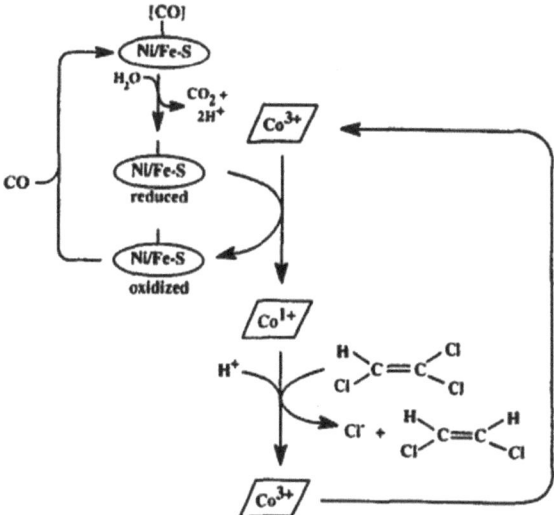

electrons to the Co/Fe–S component (Abbanat and Ferry 1991) reducing the cobalt atom to the low-potential $Co^+$ redox state ($E_m$ of $Co^{2+}/Co^+$ couple = $-515$ mV, Jablonski and Ferry 1992). In the proposed mechanism for TCE dechlorination, the Ni/Fe–S component oxidizes CO and electrons are transferred to the Co/Fe–S component reducing the Co atom of enzyme-bound factor III to the $Co^+$ redox state. The CO-reduced Ni/Fe–S component by itself does not dechlorinate TCE (Jablonski and Ferry 1992); however, the reduced enzyme-bound factor III reductively dechlorinates TCE to cis-DCE. The mechanism of reductive dechlorination by the enzyme-bound factor III is unknown, but may involve the formation of a CO–C bond as was previously shown for the dechlorination of $CCl_4$ (Krone et al. 1989) and TCE (Gantzer and Wackett 1991) with free vitamin $B_{12}$. Although the rate of dechlorination of TCE was higher when free factor III was used, the ratio and distribution of products formed was comparable to those of the CO-reduced CODH enzyme complex (Jablonski and Ferry 1992). Thus, CO-based anaerobic processes for the bioremediation of halogenated compounds are interesting prospects.

## 4.3.2 Reductive/Substitutive Dechlorination of Tetrachloromethane (CCl₄) by Clostridium Thermoaceticum and Methanobacterium Thermoautotrophicum

Chlorinated aliphatic hydrocarbons, some of which are carcinogens or mutagens, are common water pollutants. Several mono- and dihalogenated hydrocarbons are utilized by aerobic microorganisms as carbon and energy sources for growth, and some trihaloalkanes are subject to aerobic transformations (Egli et al. 1988). Microbial transformation of the industrially relevant polychlorinated hydrocarbons tetrachloromethane, tetrachloroethylene, and 1,1,1-trichloroethane, however, is catalyzed only under anaerobic conditions by undefined dehalogenation mechanisms (Vogel et al. 1987; Cook et al. 1988).

Tetrachloromethane and trichloromethane are subject to anaerobic biotransformation (Muller and Lingens 1986; Vogel et al. 1987; Cook et al. 1988). The products from $CCl_4$, which are seen in soil samples, are also found in pure cultures of several strictly anaerobic bacteria (*Methanobacterium thermoautotrophicum, A. woodii, Desulfobacterium autotrophicum,* and *Clostridium* spp.) (Egli et al. 1990; Lowe et al. 1993). Two types of reactions have been recognized: reductive, to less highly chlorinated methanes, and substitutive, to $CO_2$ and some of its transformation products (Egli et al. 1988; Galli and McCarty 1989). The ability to anaerobically transform $CCl_4$ has been correlated (Egli et al. 1988) with the presence of the acetyl-CoA pathway (Wood et al. 1986) for the degradation or synthesis of acetate (acetyl-CoA) in these organisms. Bacteria with an operative acetyl-CoA pathway contain high levels of corrinoids (Dangel et al. 1987), and it has been suggested that the latter catalyze reductive dehalogenation in anaerobic bacteria (Egli et al. 1988; Krone et al. 1989).

Total transformation of tetrachloromethane was shown by the mesophilic carboxydotrophs *D. autotrophicum* and *A. woodii*, and the thermophilic carboxydotroph *C. thermoaceticum* (Egli et al. 1988, 1990). Reduction of CCl4 by *D. autotrophicum* required 18 days of incubation, and *A. woodii* and *C. thermoaceticum* were able to degrade 80 μM $CCl_4$ completely within 3 days (Egli et al. 1988). Trichloromethane accumulated as a transient intermediate, but the only chlorinated methanes recovered at the end of the incubation period were 8 μM dichloromethane and traces of chloromethane. Therefore, 90 % of the $CCl_4$ was degraded to unknown products. Growing cultures of *A. woodii* converted 92 % of added [14]$CCl_4$ to non-halogenated products. Much of the initial radioactivity (67 %) was recovered as $CO_2$, acetate, pyruvate, and cell material; the remainder included an unknown, hydrophobic material and $CH_2Cl_2$ (Egli et al. 1988). On the basis of the reactivity of halomethanes with cobamides, and because all three organisms contain cobamides, it has been suggested that cobamides could be involved in the dehalogenation of $CCl_4$.

Egli et al. (1988) have proposed a pathway for $CCl_4$ metabolism *by A. woodii* that comprises at least two sequences. In the first sequence, corrinoid enzymes putatively catalyze $CCl_4$ reduction to trichloromethane, dichloromethane, and chloromethane. In the second sequence, a substitutive branch transforms $CCl_4$ into $CO_2$ by a series of unknown reactions, which do not cause a net change in the oxidation state of the carbon atom. $CO_2$ is then assimilated by the acetyl-CoA pathway. In the reductive pathway, $CCl_4$ and the other chlorinated methanes serve as electron acceptors. Under anaerobic conditions in reduced buffer, suspensions of *A. woodii, D. autotrophicum,* or *M. thermoautotrophicum* degraded $CCl_4$ by both reductive and substitutive mechanisms (Egli et al. 1990). The products formed included less highly chlorinated methanes and $CO_2$. Cell extracts of *A. woodii* degraded tetrachloromethane in a manner similar to that in whole cells but at a lower rate (63 vs. 140 μkat kg$^{-1}$ of protein) (Egli et al. 1990). When *M. thermoautotrophicum* or *A. woodii* was autoclaved, reductive dechlorination was partly abolished, whereas substitutive dechlorination was retained. Trichloromethane was oxidized to $CO_2$ by both native and autoclaved cells of *A. woodii*. Halomethanes are therefore degraded anaerobically by reductive, substitutive, and oxidative mechanisms (Egli et al. 1990).

## 4.3.3 Transformation of 2,4,6,-trinitrotoluene (TNT) by a Carboxydotrophic Sulfate-reducing Anaerobe

2,4,6-Trinitrotoluene (TNT) is one of the most widely used explosives. Therefore, it occurs as a pollutant of soil and ground water especially at sites of ammunition factories (Preuss et al. 1993). The fact that about 50 years after World War II at locations of former ammunition factories, still large amounts of TNT and its derivatives can be found in the soil indicates a high persistence of these compounds in natural environments (Preuss et al. 1993). The biological degradability of nitroaromatics is therefore of great interest. Attempts to find microorganisms which are able to degrade TNT under aerobic conditions within a reasonable period of time have failed so far. The anaerobic degradation of nitro compounds is usually initiated by the reduction of the nitro substituents. There are several reports on the (unspecific) reduction of aromatic or aliphatic nitro groups by anaerobes or enzymes from these organisms (O'Brien and Morris 1971; McCormick et al. 1976; Angermaier and Simon 1983; Hallas and Alexander 1983). The reduction of the first nitro substituent is catalyzed by many aerobic and anaerobic bacteria (McCormick et al. 1976; Parrish 1977; Amerkhanova and Naumova 1978; Kaplan and Kaplan 1982; Schackmann and Müller 1991). The product of this reaction, aminodinitrotoluene, is further reduced to 2,4-diamino-6-nitrotoluene (DANT) by facultative and anaerobic bacteria (McCormick et al. 1976; Naumova et al. 1989). The reduction of the latter compound to triaminotoluene was observed only with strictly anaerobic bacteria. Evidence was presented with *Veillonella alcalescens* that the reduction of DANT might be catalyzed by partially purified hydrogenase combined with a fraction containing a compound similar to or identical with ferredoxin (McCormick et al. 1976). The nitro group reduction with DANT as the substrate therefore appeared similar to that with other nitroaromatic and nitroaliphatic compounds, which was mediated by purified hydrogenase and ferredoxin or viologens (Angermaier and Simon 1983).

CO is an electron donor for the reductive transformation of 2,4,6,-trinitrotoluene by a sulfate-reducing anaerobe (Preuss et al. 1993). CO also provides electrons for the reductive carboxylation of phenols, which is the first step in the anaerobic degradation of these aromatic compounds (Knoll and Winter 1988). The extent to which CO can provide the reducing potential for other degradative and detoxification mechanisms by anaerobic microbes is unknown. CODHs have been described from phylogenetic ally diverse microbes, including species from the domains *Bacteria* and *Archaea*. The enzyme has the assorted functions of CO oxidation, acetyl-CoA synthesis, and acetyl-CoA cleavage. It is present in aerobes and anaerobes with broad metabolic capabilities. This diverse nature of CODH suggests that many more undiscovered microbes utilize the enzyme in novel pathways that could have potential in the biotechnology industry. The recent cloning of CODH genes should provide a source of DNA probes to aid in the isolation and identification of novel CODH-containing microbes.

## 4.4  Thermophilic Carboxydotrophs as Biosensors for CO Detection

Biosensors are functional analogs that are based on the direct coupling of an immobilized biologically active compound with a signal transducer and an electronic amplifier (Bănică 2012; Turner et al. 1987). The main function of a transducer is to convert the physico-chemical change in the biologically active material resulting from the interaction with the analyte into an output signal. Figure 4.20 shows a general configuration of a biosensor. A biosensor typically consists of a biorecognition component, biotransducer component, and electronic system which include a signal amplifier, processor, and display (Hierlemann and Baltes 2003; Hierlemann et al. 2003). The recognition component, often called a bioreceptor (Fig. 4.20), uses biomolecules from organisms or receptors modeled after biological systems to interact with the analyte of interest. This interaction is measured by the biotransducer which outputs a measurable signal proportional to the presence of

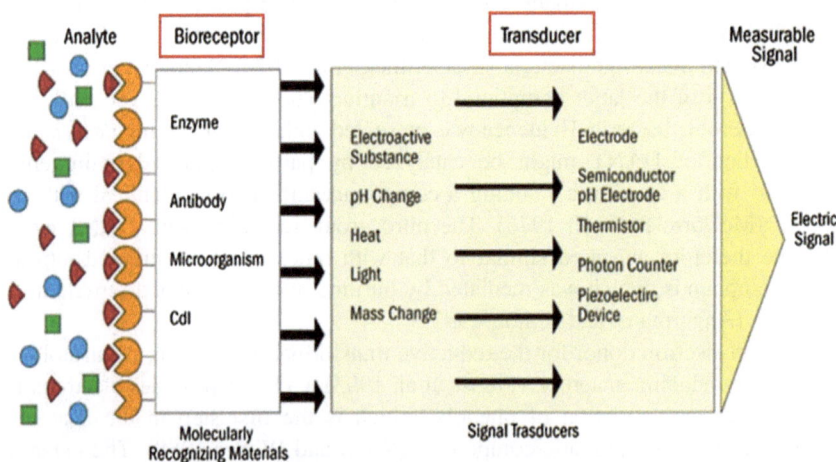

**Fig. 4.20**  Schematic of a biosensor (Chaubey and Malhotra 2002)

the target analyte in the sample. The general aim of the design of a biosensor is to enable quick, convenient testing at the point of concern or care where the sample was procured (Chaubey and Malhotra 2002). In a biosensor, the bioreceptor is designed to interact with the specific analyte of interest to produce an effect measurable by the transducer. High selectivity for the analyte among a matrix of other chemical or biological components is a key requirement of the bioreceptor. While the type of biomolecule used can vary widely, biosensors can be classified according to common types of bioreceptor interactions involving: antibody/antigen, enzymes, nucleic acids/DNA, cellular structures/cells, or biomimetic materials (Vo-Dinh and Cullum 2000). In addition to enzymes, antibodies, and nucleic acids, the biorecognition systems can also include bacteria and single cell organisms and even whole tissues of higher organisms (Chaubey and Malhotra 2002). Specific interactions between the target analyte and the complementary biorecognition layer produce a physico-chemical change which is detected and may be measured by the transducer. The transducer can take many forms depending upon the parameters being measured—electrochemical, optical, mass, and thermal changes are the most common.

Based on the type of transducer used, biosensors have been divided into optical, calorimetric, piezoelectric, and electrochemical biosensors. Optical biosensors are based on the measurement of light absorbed or emitted as a consequence of a biochemical reaction. In such a biosensor, the light waves are guided by means of optical fibers to suitable detectors (Peterson and Vurek 1984; Seitz 1987). They can be used for measurement of pH, $O_2$, or $CO_2$, etc. Many optical biosensors are based on the phenomenon of surface plasmon resonance (SPR) techniques (Zeng et al. 2014). This type of optical biosensor utilizes a property of and other materials; specifically that a thin layer of gold on a high refractive index glass surface can absorb laser light, producing electron waves (surface plasmons) on the gold surface. The production of electron waves occurs only at a specific angle and wavelength of incident light and is highly dependent on the surface of the gold, such that binding of a target analyte to a receptor on the gold surface produces a measurable signal. A commercial optical biosensor, which is the hybrid electrochemical/optical LAPS (light-addressable potentiometric sensor), was developed by the company Molecular Devices in Palo Alto, USA (Tiefenthaler 1993).

Calorimetric biosensors detect an analyte on the basis of the heat evolved due to the biochemical reaction of the analyte with a suitable enzyme (Chaubey and Malhotra 2002). The temperature changes are usually determined by means of thermistors at the entrance and exit of small packed bed columns containing immobilized enzymes within a constant temperature environment (Fig. 4.21). Under such closely controlled conditions, up to 80 % of the heat generated in the reaction may be registered as a temperature change in the sample stream. This may be simply calculated from the enthalpy change and the amount reacted. Recently, integrated circuit temperature-sensitive structures have been modified with enzymes (Chaubey and Malhotra 2002). Different substrates, enzymes, vitamins, and antigens have been determined using thermometric biosensors. The most commonly used approach in the thermal enzyme probes (Danielsson and Mosbach 1987; Weaver et al. 1976) was related to the enzyme directly attached to the thermistor.

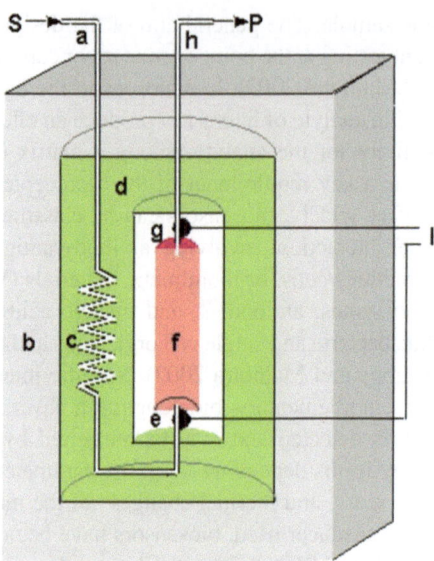

**Fig. 4.21** Schematic diagram of a calorimetric biosensor. The sample stream passes (*a*) through the outer insulated box (*b*) to the heat exchanger (*c*) within an aluminum block (*d*). From there, it flows past the reference thermistor (*e*) and into the packed bed bioreactor (*f*, 1 ml volume), which contains the biocatalyst, where the reaction occurs. The change in temperature is determined by the thermistor (*g*) and the solution passed to waste (*h*). External electronics (*l*) determines the difference in the resistance, and hence temperature, between the thermistors (http://www1.lsbu.ac.uk/water/enztech/calorimetric.html)

It was observed that the majority of the heat evolved in the enzymatic reaction was lost to the surrounding solution without being detected by thermistor resulting in the decrease in sensitivity of the biosensor.

Piezoelectric sensors utilize crystals which undergo an elastic deformation when an electrical potential is applied to them (Fig. 4.22). An alternating potential produces a standing wave in the crystal at a characteristic frequency. This frequency is highly dependent on the elastic properties of the crystal, such that if a crystal is coated with a biological recognition element, the binding of a large target analyte to a receptor will produce a change in the resonance frequency, which gives a binding signal. Adsorption of the analyte increases the mass of the crystal and alters its basic frequency of oscillation. Piezoelectric sensors are used for the measurement of ammonia, NO, carbon monoxide, hydrogen, methane, and certain organophosphorus compounds (Abad et al. 1998; Minunni et al. 1994).

Electrochemical biosensors are the most commonly used class of biosensors. They have been widely accepted in biosensing devices, can be operated in turbid media, have comparable instrumental sensitivity, and are more amenable to miniaturization (Chaubey and Malhotra 2002). Electrochemical biosensors are usually based on potentiometry and amperometry. Ion-selective electrodes (ISE), ion-selective field-effect transistors (ISFET), and pH electrodes are usually based on the oxidation

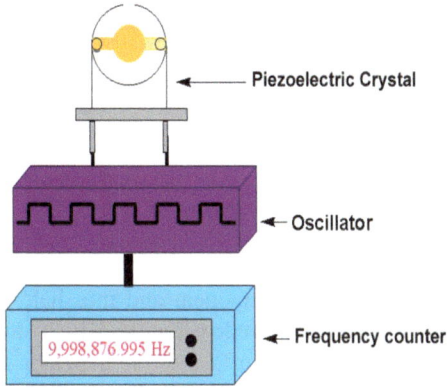

**Fig. 4.22** Schematic diagram of the antibody–antigen binding (Kumar 2000)

of the substrate/product, e.g. oxygen electrode and detection of $H_2O_2$. Depending upon the electrochemical property to be measured by a detector system, electrochemical biosensors may further be divided into conductometric (Sukeerthi and Contractor 1994), potentiometric (Papastathopoulos and Rechnitz 1975; Mascini 1995; Senillou et al. 1999; Koncki et al. 2000), and amperometric biosensors (Lindgren et al. 2000; Davis et al. 1995). Electrochemical biosensors are based on mediated or unmediated electrochemistry for electron transfer (Fig. 4.23). Ferrocene and its derivatives, ferricyanide, methylene blue, benzoquinone, N-methyl phenazine, etc., are most commonly used mediators in mediated biosensors

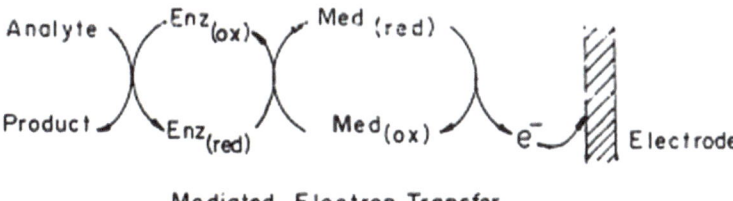

Mediated Electron Transfer

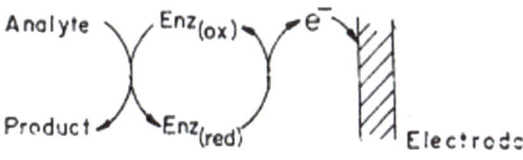

Unmediated Electron Transfer

**Fig. 4.23** The scheme of mediated and unmediated electron transfer (Chaubey and Malhotra 2002)

(Karyakin et al. 1994; Turner 1988; Kulys and Cenas 1983; Kajiya et al. 1991; Cass et al. 1984; Cenas et al. 1984; Jaffari and Turner 1997; Gregg and Heller 1991; Garjonyte et al. 2001).

## 4.4.1 Mediators

Mediators are artificial electron transferring agents that can readily participate in the redox reaction with the biological component and thus help in the rapid electron transfer. It is a low molecular weight redox couple, which shuttles electrons from the redox center of the enzyme to the surface of the indicator electrode. During the catalytic reaction, the mediator first reacts with the reduced enzyme and then diffuses to the electrode surface to undergo rapid electron transfer. A mediator is expected to be stable under required working conditions and should not participate in the side reactions during electron transfer. The mediator should be chosen in such a way that it has a lower redox potential than the other electrochemically active interferents in the sample. The redox potential of a suitable mediator should provide an appropriate potential gradient for electron transfer between enzyme's active site and electrode. The redox potential of the mediator (compared to the redox potential of enzyme active site) should be more positive for oxidative biocatalysis or more negative for reductive biocatalysis. Direct current voltammetry is a useful technique to study the properties of mediators, and it helps in selecting a suitable mediator for an amperometric biosensor (Gilmartin and Hart 1995; Nakaminami et al. 1997). Organic dyes such as methylene blue, phenazines, methyl violet, Alizarin yellow, prussian blue, thionin, azure A and azure C, toluidine blue, and inorganic redox ions such as ferricyanide have been widely used in a number of biosensors (Dubinin et al. 1991; Karyakin et al. 1994, 1995; Aoyagi et al. 1997; Molina et al. 1999; Brunetti et al. 2000). They, however, suffer from a number of problems such as poor stability and pH dependence of their redox potentials (organic dyes). The inorganic mediators have problems in that it is not easy to tune their redox potentials and solubility by the use of substituents. Recently, the use of ferrocene derivatives as redox mediators for flavo- and quinoenzymes has been worked out. Nakaminami et al. (1997) studied the electrochemical sensitivity to cholesterol in the detection system using cholesterol oxidase and these redox compounds, and the employment of MPMS or thionin allows electrochemical determination of cholesterol at a low electrode potential (0 V versus SCE) in the concentration range 0.25–0.5 mM.

## 4.4.2 Enzyme Electrodes

A simplest enzyme electrode consists of a thin layer of enzyme held in close proximity to the active surface of a transducer, a suitable reference electrode, and a circuit for measuring either by potentiometry or by amperometry (Chaubey and Malhotra 2002). While carrying out the measurements, the enzyme electrode is immersed into the analyte to be detected and the steady-state potential or current is read. In general, a logarithmic relationship is observed for the potentiometric, and a linear behavior for the amperometric electrodes. Generally, the electrocatalysts are immobilized onto an electrode surface by adsorption (Degrand and Miller 1980; Huck and Schmidt 1981; Jaegfeldt et al. 1981), polymerization (Jaegfeldt et al. 1983), or electrodeposition (Persson et al. 1993). The peak potentials for NADH electrocatalysis for different electrocatalysts were reported by Lorenzo et al. (1998). They found dramatic potential shifts as compared to the NADH oxidation at the bare glassy carbon electrode (0.7 V). For a kinetically controlled biochemical reaction catalyzed by the immobilized enzyme, the steady-state current is proportional to the initial rate of enzymatic process (Mell and Maloy 1975). In this case, a plot of I versus substrate concentration S yields a typical Michaelis–Menten-type response. A linear Lineweaver–Burke plot, 1/I versus 1/S, is the diagnostic technique for kinetic control of the electrochemical response. The response of a biosensor is typically dependent on the amount of active enzyme immobilized. A low molecular weight soluble mediator is disadvantageous as it can leach out of the electrode and be lost to the bulk solution. This may lead to a significant signal loss (Schuhmann et al. 1990) and is considered as a serious problem for in vitro applications. To overcome this, several groups have investigated the use of immobilized mediators either with enzyme in solution or with co-immobilized enzyme.

### 4.4.3 Gas Biosensors

Biosensors for gas analysis have already achieved some commercial success; an immobilized cholinesterase reactor system capable of detecting gaseous nerve poisons, such as pesticides, in the ppm range, has been developed by Goodson and Jacobs (1974). The device consists of a pad of immobilized cholinesterase held between a pair of platinum electrodes. A stream of sample air together with the enzyme substrate butyrylthiocholine iodide is passed through the pad and the hydrolysis product thiocholine iodide I is detected electrochemically. In the presence of enzyme inhibitors, no easily oxidizable thiol is formed and the cell voltage rises (Goodson and Jacobs 1974). Biosensors incorporating intact microorganisms have been described for the determination of methane (Okada et al. 1981; Karube et al. 1982), ammonia (Hikuma et al. 1980), nitrogen dioxide (Suzuki and Karabe 1982), and volatile organic compounds (VOCs) (Lee and Karube 1996; Naessens and Minh 1998). These systems require respective gases to be dissolved prior to delivery to the immobilized cells. The metabolism of the substrates is reflected by the consumption of oxygen, which is monitored using a Clark oxygen probe. The use of intact microorganism rather than the relevant purified enzymes facilitates the construction of stable devices, with outputs reported to remain steady for 10–24 days (Suzuki and Karube 1982).

The use of a purified enzyme for assay of its substrate directly in the gas phase was first demonstrated by Guilbault (1983) and represents a significant departure from conventional approaches. Guilbault (1983) has reported that the absorption may be made specific for formaldehyde by coating the crystal with a mixture of formaldehyde dehydrogenase, reduced glutathione, and NAD. The implication that the specific binding capacity of an enzyme may be exploited under dry condition has important connotations, particularly for gas analysis. Electrochemical biosensors using immobilized enzymes as a sensing element have also been developed by Dennison et al. (1995), Park et al. (1995), Hammerle and Hall (1996), and Kaisheva et al. (1997) for the detection of gases. These sensors typically demonstrated high sensitivity and selectivity. In addition, a rapid response allowing a reduction in the detection time was another feature.

## 4.4.4  Detection of CO from the Environment

There is considerable commercial interest in the potential of biosensors for specific measurement of gases, with one prime target for such development programs being CO. The CO biosensor is necessary for the following: (1) Detection of CO is necessary in mines, underground car parks, road tunnels, and various industrial situations; (2) quantification of the gas is important in combustion control systems for furnaces or engines; and (3) as a tool to provide the basis for a fire alarm (Colby et al. 1985). Many of these applications require a greater degree of specificity than is offered by conventional technology in order to minimize false alarms or readings. The implication of the gas in the emotive area of cigarette smoking is part of a broad clinical and biological interest which frequently requires accurate measurement of CO both in solution and in the gas phase.

### 4.4.4.1  Health Effects Associated with CO

CO is a by-product of the combustion of fuel and is emitted in the exhaust of gasoline, propane, or other fuel-powered equipment and engines. It is formed by the incomplete combustion of carbon-containing materials, which are present in aviation fuels. It is also present in unvented kerosene and gas space heaters; leaking chimneys and furnaces; back-drafting from furnaces, gas water heaters, wood stoves, and fireplaces; gas stoves; generators and other gasoline powered equipment; automobile exhaust from attached garages; and tobacco smoke. Incomplete oxidation during combustion in gas ranges and unvented gas or kerosene heaters may cause high concentrations of CO in indoor air. Worn or poorly adjusted and maintained combustion devices (e.g., boilers and furnaces) can be significant sources, or if the flue is improperly sized, blocked, disconnected, or is leaking. Auto, truck, or bus exhaust from attached garages, nearby roads, or parking areas can also be a source (see www.epa.gov/airquality/carbonmonoxide). CO is a hidden danger because it is a colorless and odorless gas. Exposure to CO can cause harmful health effects depending on the air concentration and duration of exposure. CO is an asphyxiant in humans, where inhalation causes tissue hypoxia by preventing the blood from carrying sufficient oxygen. Acute CO poisoning is associated with headache, dizziness, fatigue, nausea, and at elevated doses, neurological damage, and death. Higher acute exposure or chronic exposures can also affect the heart, particularly in those with cardiovascular disease. Exposure to CO can result in individuals becoming confused or incapacitated before they are able to leave the contaminated environment. When this occurs in an aircraft, the end result could quite possibly be an accident. Table 4.7 lists the symptoms that can be expected based on the amount of CO in the area and as a function of duration of exposure (Tierney 2004). The current Occupational Safety and Health Administration (OSHA) permissible exposure limit (PEL) for carbon monoxide is 50 parts per million (ppm) parts of air [55 milligrams per cubic meter (mg m$^{-3}$)] as an 8-h time-weighted average (TWA) concentration (Benignus et al. (1979). The symptoms of mild headache, nausea, and fatigue can occur at 200 ppm between 2 and 3 h of exposure, where an increasing magnitude of exposure for shorter periods of time results in similar symptoms. At extreme exposure (12,800 ppm), it only takes 1–3 min to cause death (Tierney 2004).

Typical symptoms as a function of CO exposure in terms of concentration in the blood, which is shown in Table 4.8. Slight headaches begin at 10 % blood content of CO, drowsiness begins at around 20 % blood content of CO, and blurring of vision is present starting around 30 % blood content of CO. Unconsciousness and death can occur when the amount of CO is more than 50 % in the blood.

**Table 4.7** Symptoms resulting from CO exposure (Tierney 2004; Occupational Safety and Health Standards 1997)

| CO (ppm) | Time | Exposure of symptoms |
|---|---|---|
| 50 | 8 h | Maximum exposure allowed by the Occupational Safety and Health Administration over an 8-h period |
| 200 | 2–3 h | Mild headache, nausea, fatigue |
| 400 | 1–2 h | Serious headache, life threatening after 3 h |
| 800 | 45 min | Dizziness, nausea, unconscious within 2 h, death within 2–3 h |
| 1,600 | 20 min | Headache, dizziness, nausea, death within 1 h |
| 3,200 | 5–10 min | Headache, dizziness, nausea, death within 1 h |
| 6,400 | 1–2 min | Headache, dizziness, nausea, death within 25–30 min |
| 12,800 | 1–3 min | Death |

ppm = parts per million

**Table 4.8** Percentage of CO in the blood and possible symptom (Tierney 2004)

| CO in blood (%) | Typical of symptoms |
|---|---|
| <10 | None |
| 10–20 | Slight headache |
| 21–30 | Headache, slight increase in respirations, drowsiness |
| 31–40 | Headache, impaired judgment, shortness of breath, increasing drowsiness, blurring of vision |
| 41–50 | Pounding headache, confusion, marked shortness of breath, marked drowsiness, increasing blurred vision |
| >50 | Unconsciousness, eventual death if victim is not removed from the source of CO |

### 4.4.4.2 CO Biosensor

The diversity of situations in which CO measurement is required is mirrored by the range of conventional techniques adopted. In situations where expense and size are not an issue, gash chromatography and infrared absorption have been used. Mass spectrophotometer has also been used in the past; however, problems were encountered in the presence of dinitrogen gas, which has the same mass number as CO (McLafferty 1993). A limited range of low-cost miniature CO gas detectors are commercially available, which operate either by the direct electrochemical oxidation of CO of by absorption at coated semiconductor devices (Watson 1984; Azad et al. 1992). When analyzing real samples, however, the most important limitation of these devices is the lack of selectivity and interference in $N_2O$, $NO_2$, and $H_2S$ (Wetchakun et al. 2011).

A prerequisite for a successful biosensor is the identification of an appropriate biological component. The ability of CO to bind the terminal oxidases of aerobic organisms and to inhibit the respiration of most of them provides the basis of one approach. Intact microorganisms have been incorporated into biosensors based on secondary transducers, such as oxygen electrodes (Divies 1975), and on mediated electron transfer (Turner et al. 1983). In addition, cytochrome oxidase has been coupled to an electrode via cytochrome $c$ (Hill et al. 1981). A more flexible approach to CO quantification required the specific metabolism of the gas. CO-oxidizing bacteria posses CO oxidoreductases. This group of enzymes has been isolated in various carboxydotrophic microorganisms including the thermophilic *Pseudomonas thermocarboxydovorans* (Bell et al. 1985). The characteristics of the CO oxidoreductase from *Pseudomonas carboxydovorans* are particularly suited for use in an enzyme-based sensor, offering greater stability (half-life of several days at room temperature) and higher affinity for C ($K_m$ for CO = 0.6 µM) O than the enzymes from the carboxydotrophic bacteria (Bell et al. 1985).

CO oxidoreductases have been observed in several mesophilic carboxydotrophs (Cypionka et al. 1980) such as *P. carboxydovorans* (Meyer and Schlegel 1980) and *Pseudomonas carboxydohydrogena* (Kim and Henegan 1981). Chemical and spectral analyses have shown the CO oxidoreductase from *P. carboxydovorans* to be an iron–sulfur molybdenoflavoprotein containing two molecules of FAD, eight atoms of iron, eight of acid-labile sulfide, and probably two of molybdenum (Meyer 1982). Electron paramagnetic resonance (EPR) spectroscopy revealed the enzyme from *P. carboxydovorans* and *P. carboxydohydrogena* to contain Mo (V) atoms in two different chemical environments and two different iron–sulfur centers (Bray et al. 1983). The properties CO oxidoreductase from the thermophilic *P. thermocarboxydovorans* resembles the analogous enzyme from mesophilic bacteria (Table 4.9). Nevertheless, CO oxidoreductase from *P. thermocarboxydovorans* shows the following unique properties: (1) the enzyme is heat-stable with an optimum temperature in the usual assay of 80 °C; (2) the enzyme has a very low apparent $K_m$ for CO (0.5), some 100-fold lower than published values for other CO oxidoreductases (Meyer and Schlegel 1980; Kim and Hegeman 1981); and (3) the enzyme can use both horse heart cytochrome *c* and potassium ferricyanide as electron acceptors (Bell et al. 1985). The natural electron acceptor of *P. thermocarboxydovorans* is ubiquinone (Bell et al. 1985). CODH from anaerobic acetogenic carboxydotrophs such as *C. thermoaceticum* and *Acetogenic woodii* has also been described (Drake et al. 1980; Ragsdale et al. 1983). However, their CODHs contain nickel and use electron acceptors of much lower potential such as viologen dyes, ferredoxin, and rubredoxin.

Figure 4.24 shows the scheme for biosensor construction with redox enzyme such as CO oxidoreductase. Alternatively, low molecular weight mediators such as phenazine ethosulfate (PES) may be used in placed of cytochrome *c* (Fig. 4.25) to couple the oxidation reaction to an electrode (Turner et al. 1982; Davis et al. 1983).

**Table 4.9** Kinetic and molecular properties of the CO oxidoreductase from *Pseudomonas thermocarboxydovorans* (Bell et al. 1985)

| Properties | |
| --- | --- |
| Optimum pH | 7.5 |
| Temperature for maximum activity | 80 °C |
| $K_m$ for CO | 0.6 μM |
| Inhibitors | Acetylene, cyanide, methanol, 8-hydroxyquinone, iodoacetate, *p*-hydroxymercuribenzoate |
| Electron acceptors | Methylene blue, ferricyanide, thionin, phenazine ethosulfate/methosulfate, cytochrome *c*, 2,6-dichlorophenolindophenol |
| $K_m$ for phenazine ethosulfate | 3.8 μM |
| Molecular weight (by gel filtration) | 310,000 |
| Molecular weight (by sedimentation velocity) | 230,000 |
| Fe content | 6.9 g atom mol$^{-1}$[a] |
| Acid-labile sulfide | 6.9 g atom mol$^{-1}$[a] |
| Mo content | 0.7 g atom mol$^{-1}$[a] |
| Flavin | 1.8[a] |
| Absorption maximum | 340 and 420 nm |

[a] Calculated assuming a molecular weight of 270,000

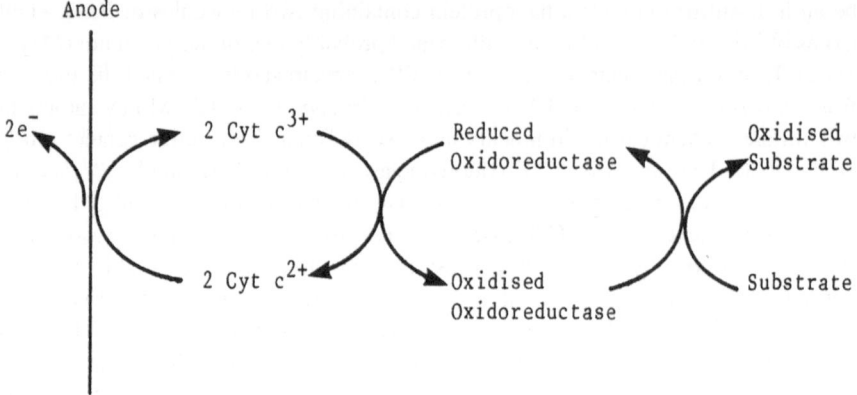

**Fig. 4.24** Scheme representing biosensor construction with redox enzyme (Hill et al. 1981)

A ferrocenium ion has also been used as a substitute mediator of electron transfer between CO oxidoreductase and a graphite electrode. This class of mediator combines well-behaved electrochemistry and insensitivity to oxygen with a choice of physical and chemical properties suitable for enzyme-based sensors (Cass et al. 1984). The kinetics of the homogeneous reaction between ferrocene monocarboxylic acid and CO oxidoreductase were studied by using DC cyclic voltammetry. The effect adding CO oxidoreductase on the transversable electrochemistry of ferrocene monocarboxylic acid in the presence of substrate is shown in Fig. 4.26. Enzyme was added to final concentrations of 10–100 μM and the enhanced anodic current obtained was recorded as a function of scan rate (Turner et al. 1985). Although the cyclic voltammetry experiments showed no indication of direct electrochemistry of either CO oxidoreductase or substrate, the enhanced anodic current obtained in the presence of the ferrocenium ($Fecp_2R^+$) was indicative of the sequence shown in Fig. 4.27.

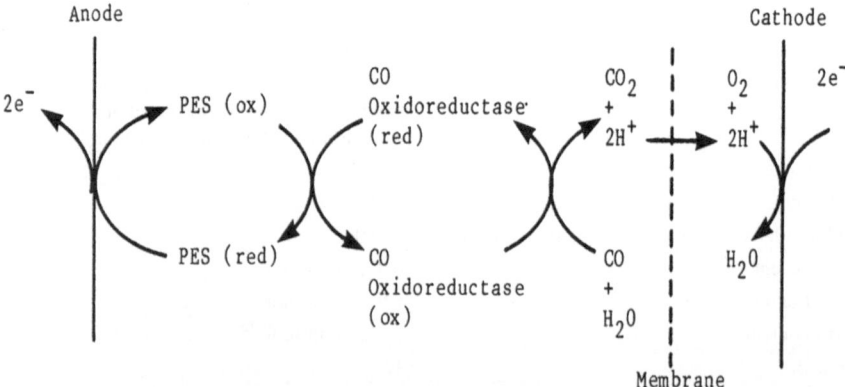

**Fig. 4.25** Reactions on CO fuel cell (Turner et al. 1982)

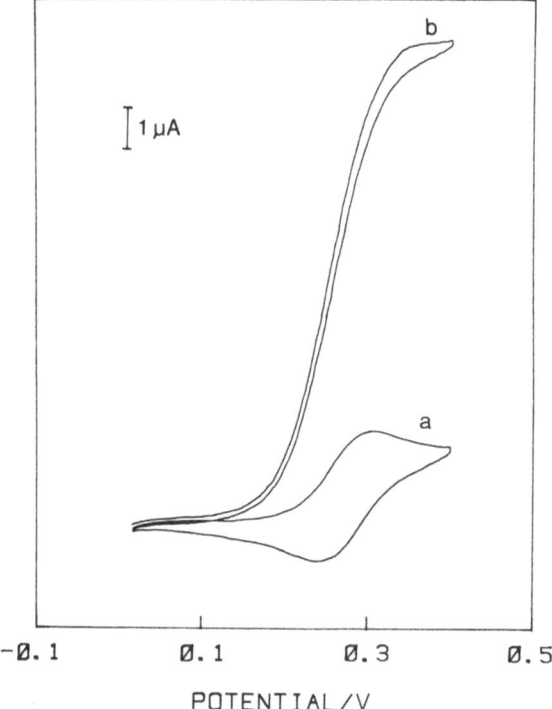

**Fig. 4.26** Direct current cyclic voltammogram of ferrocene monocarboxylic acid in argon-saturated Tris-HCl buffer containing **a** either no additions, CO or CO oxidoreductase and **b** Co and Co oxidoreductase (Cass et al. 1984)

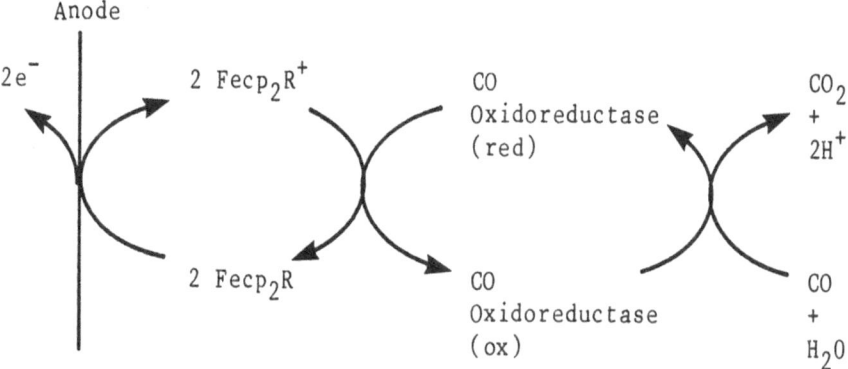

**Fig. 4.27** Sequence indicated by enhanced anodic current observed in cyclic voltammetry experiments in the presence of ferrocenium ion (Bell et al. 1985)

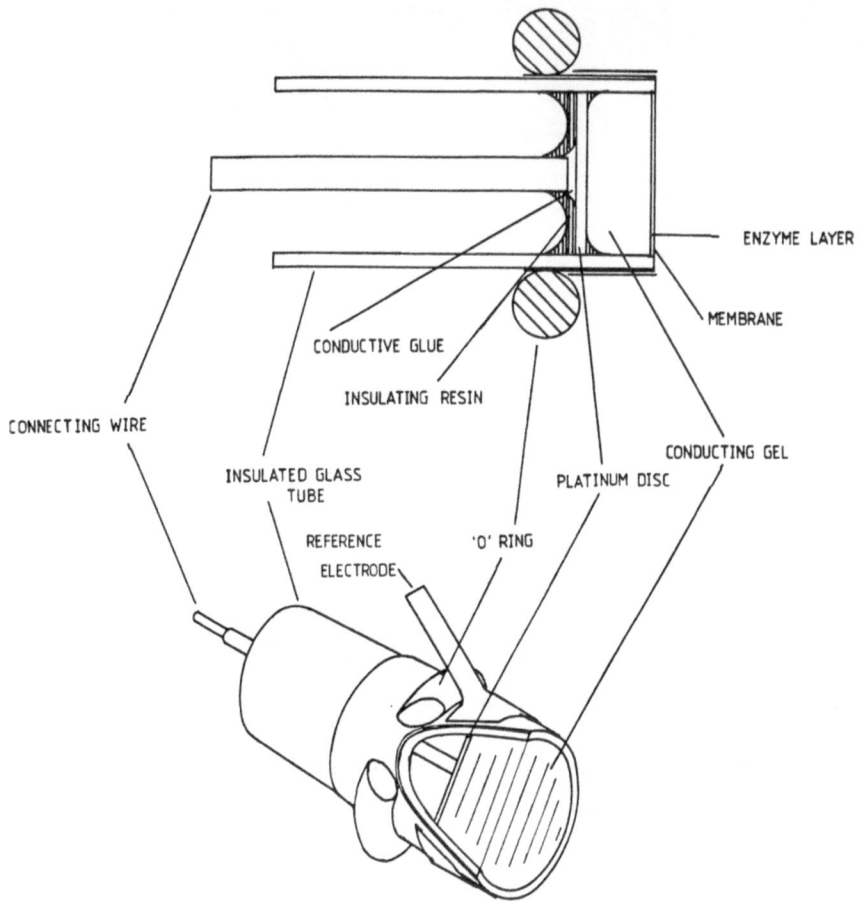

ENZYME LAYER

MEMBRANE

CONDUCTIVE GLUE

INSULATING RESIN

CONNECTING WIRE

CONDUCTING GEL

INSULATED GLASS
TUBE

PLATINUM DISC

REFERENCE         'O' RING
ELECTRODE

**Fig. 4.28** An enzyme-based CO probe (Colby et al. 1985)

Figure 4.28 illustrates an enzyme-based CO sensor. The working electrode was constructed from a platinum disk coated with a carbon paste containing the virtually insoluble ferrocene derivative 1,1′–dimethylferrocene as mediator (Turner et al. 1985). CO oxidoreductase is retained at the electrode surface by using a gas-permeable membrane. A cylindrical piece of silver foil is used as pseudo-reference electrode. A potential of +150 mV versus Ag/AgCl is applied to the enzyme electrodes with a BBC microcomputer equipped with a programmable interface (Fig 4.29). The amperometric responses of the electrodes are recorded by a computer. Probes respond very rapidly to CO both in solution and as a gas, reaching a steady-state current in less than 15 s. The current obtained is directly proportional to the aqueous CO concentration up to 60 μM, which is well above the $K_m$ of the enzyme in the homogenous solution ($K_m$ for CO = 0.6 μM) (Turner et al. 1985).

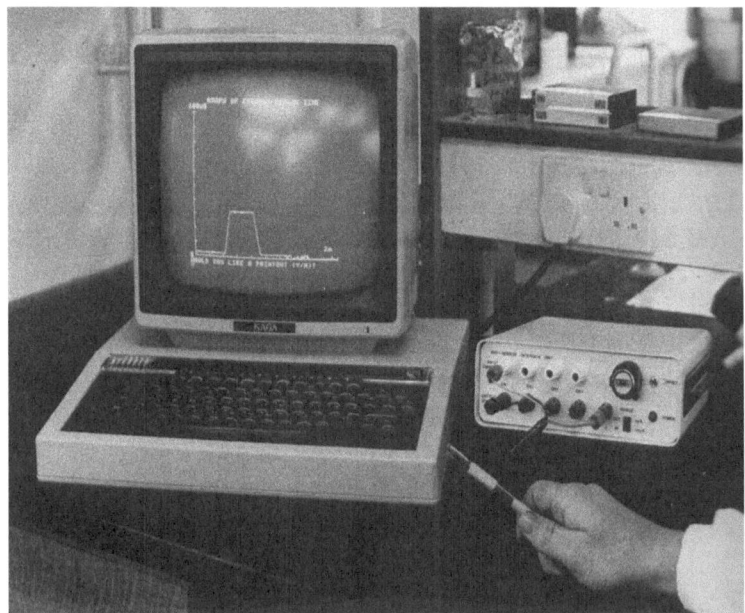

**Fig. 4.29** An enzyme-based CO probe connected to a BBC microcomputer via a programmable interface (Bell et al. 1985)

The probes are relatively stable with the current decreasing by approximately 12 % per hour on continual operation (Turner et al. 1985).

The substantial currents produced by this type of amperometric biosensor coupled with its potential for mass production from simple materials suggest that this is a promising route to a cheap miniature CO sensor with high specificity. It is a simple, cheap, and continuous monitoring device that has been demonstrated based on the highly specific CO-oxidizing enzyme. The enzyme is cheated from delivering electrons to its natural acceptor, oxygen. Instead, electrons are delivered to a current measuring device via mediators such as cytochrome $c$ or ferrocene. The current flowing gives a highly specific, rapid, and sensitive measure of CO in the ppb to ppm range. The sensitivity and specificity of a CO biosensor operated in the laboratory already lie within a useful range for the applications mentioned above. However, designs capable of operating in an industrial environment on dry gas samples will be more difficult to achieve. Probably the biggest problem is one of useful life. Devices with a long storage-life, but limited useful life, can be envisaged, which would have a limited application for one-shot testing. The really robust maintenance-free industrial CO biosensor, however, awaits considerable improvement in the stability of the system.

# Chapter 5
# Conclusions

Much is now known about carboxydotrophs. In the near future, carboxydotrophs are going to play an increasingly important role in various industries and environmental biotechnology as a source for new enzymes and processes that will occur without pre-treatment to less extreme conditions. Recent research developments further highlight the tremendous potential of such microorganisms, opening the way to a whole range of new applications.

Microbial production of electricity may become an important form of bioenergy in future because MFCs offer the possibility of extracting electric current from a wide range of soluble or dissolved complex organic wastes and renewable biomass. Thus, electricity generation in microbial fuel cells (MFCs) has been a subject of significant research efforts. While MFC application for electricity production from a variety of organic sources has been demonstrated, very little research on electricity production from carbon monoxide and synthesis gas (syngas) in an MFC has been reported. An MFC-based process of syngas conversion to electricity offers a number of advantages such as high Coulombic efficiency and biocatalytic activity in the presence of carbon monoxide and sulfur components. However, the optimization of biological and engineering components is necessary prior to application of the design. In comparison to the mesophilic counterparts, the thermophilic MFCs could demonstrate increased metabolic and current production rates as indicated by other complex bioprocesses at elevated temperatures. Though the current and power yields are relatively low at present, it is expected that with improvements in technology and knowledge about these unique systems, the amount of electric current (and electric power) which can be extracted from these systems will increase tremendously providing a sustainable way of directly converting syngas to useful energy.

Biofuels are a promising source of sustainable energy. Biomass can be converted to synthesis gas (syngas) through gasification and transformed to fuels using carboxydotrophic microorganisms that can convert the CO, $H_2$, and $CO_2$ in synthesis gas to fuels such as ethanol, butanol, and hydrogen. The discovery of novel thermophilic carboxydotrophs capable of higher product yield, as well as metabolic

© The Author(s) 2014
S.M. Tiquia-Arashiro, *Thermophilic Carboxydotrophs and their Applications in Biotechnology*, Extremophilic Bacteria, DOI 10.1007/978-3-319-11873-4_5

engineering of existing microbial catalysts, makes this technology a viable option for reducing our dependency on fossil fuels. The syngas fermenting thermophilic carboxydotrophic microorganisms reported here posses advantageous characteristics for biofuel production and hold potential for future engineering efforts. Unfortunately, few thermophilic carboxydotrophs that produce biofuels from syngas have been isolated. Thermophilic carboxydotrophs may have an advantage as opposed to mesophilic microorganisms as less cooling of the syngas is required before it is introduced into the bioreactor. Moreover, higher temperatures cab lead o higher conversion rates and benefit separation of the product by distillation. If conditions are chosen properly, it is possible to isolate thermophilic carboxydotrophs that produce biofuels (e.g. ethanol or butanol).

Thermophilic carboxydotrophs possess substantial potential for the degradation of environmental pollutants. However, knowledge of thermophilic degradation pathways is still very limited and the properties of the enzymes involved are largely unknown. This is an interesting area for future research, which needs to be exploited. New degradation pathways with new reactions catalysed by novel enzymes will be discovered, which will lead to novel biotechnological processes.

One of the most rapid and substantial returns for research investment in new areas of biotechnology is liable to come from the field of biosensors. A smaller but significant market exists for process control and environmental sensors (e.g. CO biosensor). Detection of CO is required in mines, underground car parks, road tunnels and various industrial situations. Many of these applications require a greater degree of specificity than is offered by conventional technology in order to minimize false alarms or readings. A further consideration is the potential market that may be available for very low-cost devices. Most of the work on CO biosensors has been carried out with dissolved CO, although sensors have been shown to function in a gas stream. The development of this principle is required to produce manufacturable sensors with realistic performance characteristics.

# References

Abad JM, Pariente F, Hernandez L, Abruna HD, Lorenzo E (1998) Determination of organophosphorus and carbamate pesticides using a piezoelectric biosensors. Anal Chem 70:2848–2855

Abbanat DR, Ferry JG (1991) Resolution of component proteins in an enzyme complex from *Methanosarcina thermophila* catalyzing the synthesis or cleavage of acetyl-CoA. Proc Natl Acad Sci USA 88:3272–3276

Abrini J, Naveau H, Nyns EJ (1994) *Clostridium autoethanogenum*, sp. nov., an anaerobic bacterium that produces ethanol from carbon monoxide. Arch Microbiol 161:345–351

Abubackar HN, Veiga MC, Kennes C, Coruña L (2011) Biological conversion of carbon monoxide: rich syngas or waste gases to bioethanol. Biofuels Bioprod Biorefin 5:93–114

Adams MW (1990) The structure and mechanism of iron-hydrogenases. Biochim Biophys Acta 1020:115–145

Adamse AD (1980) New isolation of *Clostridium aceticum* (Wieringa). Antonie Van Leeuwenhoek J Microbiol 46:523–531

Afting C, Kremmer E, Brucker C, Hochheimer A, Thauer RK (2000) Regulation of the synthesis of H$_2$-forming methylenetetrahydromethanopterin dehydrogenase (Hmd) and of HmdII and HmdIII in *Methanothermobacter marburgensis*. Arch Microbiol 174:225–232

Afting C, Hochheimer A, Thauer RK (1998) Function of H$_2$-forming methylenetetrahydromethanopterin dehydrogenase from *Methanobacterium thermoautotrophicum* in coenzyme F$_{420}$ with H$_2$. Arch Microbiol 169:206–210

Ahmed A, Lewis RL (2007) Fermentation of biomass generated synthesis gas: effects of nitric oxide. Biotechnol Bioeng 97:1080–1086

Ahmed A, Cateni BG, Huhnke RL, Lewis SR (2006) Effects of biomass-generated producer gas constituents on cell growth, product distribution and hydrogenase activity of *Clostridium carboxidivorans* P7T. Biomass Bioener 30:665–672

Alexander M (1999) Biodegradation and bioremediation. Gulf Professional Publishing, Texas. p 453

Allen TD, Caldwell ME, Lawson Pa, Huhnke RL, Tanner RS (2010) Alkalibaculum bacchi gen. nov., sp. a CO-oxidizing, ethanol-producing acetogen isolated from livestock-impacted soil. Inter J Syst Evol Microbiol 60:2483–2489

Alves JI, van Gelder AH, Alves MM, Sousa DZ, Plugge CM (2013) *Moorella stamsii* sp. nov., a new anaerobic thermophilic hydrogenogenic carboxydotroph isolated from digester sludge. Int J Syst Evol Microbiol 63:4072–4076

Amerkhanova NN, Naumova RP (1978) 2,4,6-Trinitrotoluene as a source of nutrition for bacteria. Microblologiya 47:393–395

Andreese Jr, Gottscha G, Schlegel HG (1970) *Clostridium formicoaceticumnov* spec-isolation, description and distinction from *C aceticum* and *C thermoaceticum* Archiv Fur Mikrobiologie 72:154

© The Author(s) 2014
S.M. Tiquia-Arashiro, *Thermophilic Carboxydotrophs and their Applications in Biotechnology*, Extremophilic Bacteria, DOI 10.1007/978-3-319-11873-4

Angermaier L, Simon H (1983) On the reduction of aliphatic and aromatic nitro compounds by clostridia, the role of ferredoxin and its stabilization. Hoppe-Seyler's Z Physiol Chem 364:961–975

Aoyagi T, Nakamura A, Ikeda H, Ikeda T, Mihara H, Ueno A (1997) Alizarine yellow-modified-cyclodextrine as a guest responsive absorption change sensor. Anal Chem 69:659–663

Argyros DA, Tripathi S, Barrett TF, Rogers SR, Feinberg LF, Olson DG, Foden JM, Miller BB, Lynd LR, Hogsett D, Caiazza NC (2011) High ethanol titers from cellulose by using metabolically engineered thermophilic, anaerobic microbes. Appl Environ Microbial 77:8288–8294

Attaway HH, Camper NC, Paynter MJB (1982) Anareobic microbial degradation of diuron by pond sediment. Pestic Biochem Physiol 17:96–101

Azad AM, Mhaisalkar SG, Birkefeld LD, Akbar SA, Goto KS (1992) Behavior of a new ZrO2-MoO3 sensor for carbon monoxide detection. J Electrochem Soc 139:2913–2920

Bae SS, Kim YJ, Yang SH, Lim JK, Jeon JH, Lee HS, Kang SG, Kim SJ, Lee JH (2006) *Thermoccoccus onnurineus* sp. nov., a hyperthermophilic archaeon isolated from a deep-sea hydrothermal vent area at the PACMANUS field. J Microbiol Biotechnol 16:1826–1831

Bagramyan K, Trchounian A (2003) Structural and functional features of formate hydrogen lyase, an enzyme of mixed-acid fermentation from *Escherichia coli*. Biochemistry-Moscow 68:1159–1170

Baker KH, Herson DS (1994) Bioremediation. McGraw-Hill, New York, p 375

Balch WE, Schoberth S, Tanner RS, Wolfe RS (1977) *Acetobacterium*, a new Genus of hydrogen-oxidizing, carbon dioxide-reducing, anaerobic bacteria. Int J Syst Bacteriol 27:355–361

Balk M, Weijma J, Friedrich MW, Stams AJM (2003) Methanol utilization by a novel thermophilic homoacetogenic bacterium, *Moorella mulderi* sp. nov., isolated from a bioreactor. Arch Microbial 179:315–320

Balk M, van Gelder T, Weelink SA, Stams AJM (2008) (Per)chlorate reduction by the thermophilic bacterium *Moorella perchloratireducens* sp. nov., isolated from underground gas storage. Appl Environ Microbiol 74:403–409

Balk M, Heilig HG, van Eekert MH, Stams AJ, Rijpstra IC, Sinninghe-Damsté JS, de Vos WM, Kengen SW (2009) Isolation and characterization of a new CO-utilizing strain, *Thermoanaerobacter thermohydrosulfuricus* subsp. carboxydovorans, isolated from a geothermal spring in Turkey. Extremophiles 13:885–894

Bănică FG (2012) Chemical sensors and biosensors: fundamentals and applications. Wiley, Chichester, p 576

Barik S, Prieto S, Harrison S, Clausen E, Gaddy J (1988) Biological production of alcohols from coal through indirect liquefaction. Appl Biochem Biotechnol 18:363–378

Barik S, Prieto S, Harrison SB, Clausen EC, Gaddy JL (1990) Biological production of ethanol from coal synthesis gas. In: Wise DL (ed) Bioprocessing and biotreatment of coal, 1st edn. Marcel Dekker, New York, pp 131–154

Bartholomew GW, Alexander M (1982) Microorganisms responsible for the oxidation of carbon monoxide in soil. Environ Sci Technol 16:300–301

Baschuk JJ, Li X (2001) Carbon monoxide poisoning of proton exchange membrane fuel cells. Int J Energ Res 25:695–713

Bedard C, Knowles R (1989) Physiology, biochemistry, and specific inhibitors of $CH_4$, $NH_4$, and CO oxidation by methanotrophs and nitrifiers. Microbiol Rev 53:68–84

Bell JM, William E, Colby J (1985) Carbon monoxide oxidoreductases from thermophilic carboxydobacteria. In: Poole RK, Dow CS (eds) Microbial gas metabolism: mechanistic, metabolic and biotechnological aspects. special publications of the society for general microbiology, Academic Press, New York, pp 153–159

Benemann JR (1996) Hydrogen biotechnology: progress and prospects. Nat Biotechnol 14:1101–1103

Benemann JR (1999) The technology of biohydrogen. In: Zaborsky OR, Benemann JR, Matsunaga T, Miyake J, San Pietro A (eds) BioHydrogen. Springer, New York, pp 19–30

Benignus V, Grant L, Mckee D, Raub J (1979) Revised evaluation of health effects associated with carbon monoxide exposure: an addendum to the 1979 EPA air quality criteria document for carbon monoxide. United States Environmental Protection Agency, Washington, DC, EPA/ 600/8-83/033F (NTIS PB85103471)

Bennett G (1995) The central metabolic pathway from acetyl-CoA to butyryl-CoA in *Clostridium acetobutylicum*. FEMS Microbiol Rev 17:241–249

Bennett B, Lemon BJ, Peters JW (2000) Reversible carbon monoxide binding and inhibition at the active site of te Fe-only hydrogenase. Biochemistry 39:7455–7460

Berkessel A, Thauer RK (1995) On the mechanism of catalysis by a metal-free hydrogenase from methanogenic archaea - enzymatic transformation of $H_2$ without a metal and its analogy to the chemistry of alkanes in superacidic solution. Angew Chem Int Ed 34:2247–2250

Bertoldo C, Antranikian G (2006) The order *Thermococcales*. In: Dworkin M, Falkow S, Rosenberg H, Schleifer KH, Stackebrandt E (eds) The prokaryotes—a handbook on the biology of bacteria. Springer, New York, pp 69–81

Bhatnagar L, Li SP, Jain MK, Zeikus JG (1989) Growth of methanogenic and acidogenic bacteria with pentachlorophenol as co-substrate. In: Lewandowski G, Armenante P, Baltis B (eds) Biotechnology applications in hazardous waste treatment. Engineering Foundation, New York, pp 383–393

Biegel E, Schmidt S, Müller V (2009) Genetic, immunological and biochemical evidence for a Rnf complex in the acetogen *Acetobacterium woodii*. Environ Microbial 11:1438–1443

Blumer-Schuette SE, Kataeva I, Westpheling J, Adams MW, Kelly RM (2008) Extremely thermophilic microorganisms for biomass conversion: status and prospects. Curr Opin Biotechnol 19:210–217

Boateng AA, Banowetz GM, Steiner JJ, Barton TF, Taylor DG, Hicks KB, El-Nashaar H, Sethi VK (2007) Gasification of Kentucky bluegrass (*Poa pratensis* L.) straw in a farm-scale reactor. Biomass Bioener 31:153–161

Boerrigter H, Den Uil H, Calis HP (2002) Green diesel from biomass via fischer–tropsch synthesis: new insights in gas cleaning and process design. In: Proceedings of pyrolysis and gasification of biomass and waste, expert meeting; Strasbourg, France, (September 30 October 1)

Bomar M, Hippe H, Schink B (1991) Lithotrophic growth and hydrogen metabolism by *Clostridium magnum*. FEMS Microbiol Lett 83:347–350

Bonam D, Ludden PW (1987) Purification and characterization of carbon monoxide dehydrogenase, a nickel, zinc, iron-sulfur protein, from *Rhodospirillum rubrum*. J Biol Chem 262:2980–2987

Bonch-Osmolovskaya EA, Miroshnichenko ML, Slobodkin AI, Sokolova TG, Karpov GA, Kostrikina NA, Zavarzina DG, Prokof'eva MI, Rusanov II, Pimenov NV (1999) Biodiversity of anaerobic lithotrophic prokaryotes in terrestrial hot springs of Kamchatka. Mikrobiologiya 68:343–351

Bond DR, Lovley DR (2003) Electricity production by *Geobacter sulfurreducens* attached to electrodes. Appl Environ Microbiol 69:1548–1555

Boopathy R (2000) Factors limiting bioremediation technologies. Biores Technol 74:63–67

Bowles LK, Ellefson WL (1985) Effects of butanol on *Clostridium acetobutylicum*. Appl Environ Microbiol 50:1165–1170

Boyd SA, Shelton DR (1984) Anaerobic biodegradation of chlorophenols in fresh and acclimated sludge. Appl Environ Microbiol 47:272–277

Boyd SA, Shelton DR, Berry D, Tiedje JM (1983) Anaerobic biodegradation of phenolic compounds in digested sludge. Appl Environ Microbiol 46:50–54

Boyles D. (1984) Bioenergy technology—thermodynamics and costs. Wiley, New York, p 158

Braun M, Mayer F, Gottschalk G (1981) *Clostridium aceticum* (Wieringa), A microorganism producing acetic acid from molecular hydrogen and carbon dioxide. Arch Microbiol 128:288–293

Bray RC, George GN, Lange R, Meyer O (1983) Studies by e.p.r. spectroscopy of carbon monoxide oxidases from *Pseudomonas carboxydovorans* and *Pseudomonas carboxydohydrogena*. Biochem J 211:687–694

Bredwell MD, Srivastava P, Worden RM (1999) Reactor design issues for synthesis-gas fermentations. Biotechnol Prog 15:834–844

Bridgwater AV (1994) Catalysis in thermal biomass conversion. Appl Catal A 116:5–47

Brown RC (2003) Biorenewable resources: engineering new products from agriculture. Blackwell Publishing, Ames, Iowa, p 286

Brogren C, Karlsson HT, Bjerle I (1997) Absorption of NO in an alkaline solution of $KMnO_4$. Chem Engin Technol 20:396–402

Bruant G, Lévesque MJ, Peter C, Guiot SR, Masson L (2010) Genomic analysis of carbon monoxide utilization and butanol production by *Clostridium carboxidivorans* strain P7. PloS One 5:e13033

Brunetti B, Ugo P, Moretto LM, Martin CR (2000) Electrochemistry of phenothiazine and methyl viologen biosensor electron transfer mediators at nanoelectrode ensembles. J Electroanal Chem 491:166–174

Bullister JL, Guinasso NL, Schink DR (1982) Dissolved hydrogen, carbon-monoxide, and methane at the Cepex site. J Geophys Res-Oceans Atmos 87:2022–2034

Buschhorn H, Dürre P, Gottschalk G (1989) Production and Utilization of Ethanol by the Homoacetogen *Acetobacterium woodii*. Appl Environ Microbial 55:1835–1840

Buurman G, Shima S, Thauer RK (2000) The metal-free hydrogenase from methanogenic archaea: evidence for a bound cofactor. FEBS Lett 485:200–204

Byrne-Bailey KG, Wrighton KC, Melnyk RA, Agbo P, Hazen TC, Coates JD (2010) Complete genome sequence of the electricity-producing *Thermincola potens* strain JR. J Bacteriol 192:4078–4079

Carere CR, Sparling R, Cicek N, Levin DB (2008) Third generation biofuels via direct cellulose fermentation. Int J Mol Sci 9:1342–1360

Cartman S, Minton N (2010) Method of double crossover homologous recombination in *Clostridia*. Patent WO/2010/084349, 29 July

Carver SM, Vuorirantab P, Tuovinena OH (2011) A thermophilic microbial fuel cell design. J Power Sources 196:3757–3760

Cass AE, Davis G, Francis GD, Hill HA, Aston WJ, Higgins IJ, Plotkin EV, Scott LD, Turner AP (1984) Ferrocene-mediated enzyme electrode for amperometric determination of glucose. Anal Chem 56:667–671

Chappelle EW (1962) Carbon monoxide oxidation by algae. Biochim Biophys Acta 62:45–62

Charlier D, Droogmans L (2005) Microbial life at high temperatures, the challenges, the strategies. Cell Mol Life Sci 62:2974–2984

Charpentier JC (1981) Mass-transfer rates in gas–liquid absorbers and reactors. In: Drew TB, Cokelet GR, Hoopes HW Jr, Vermeulen T (eds) Advances in chemical engineering, vol 11. Academic Publisher, New York, pp 1–33

Chen SY, Pan SH (2010) Simultaneous metal leaching and sludge digestion by thermophilic microorganisms: effect of solids content. J Hazard Mater 179:340–347

Cho K, Zholi A, Frabutt D, Flood M, Floyd D, Tiquia SM (2012) Linking bacterial diversity and geochemistry of uranium-contaminated groundwater. Environ Technol 33:1629–1640

Choi Y (2004) Construction of microbial fuel cells using thermophilic microorganisms, *Bacillus licheniformis* and *Bacillus thermoglucosidasius*. Bull Korean Chem Soc 25:813–818

Chisti Y, Kasper M, Moo-Young M (1990) Mass transfer in external loop airlift bioreactors using static mixers. Can J Chem Eng 68:45–50

Chu H, Chien TW, Li SY (2001) Simultaneous absorption of $SO_2$ and NO from flue gas with $KMnO_4$/NaOH solutions. Sci Tot Environ 275:127–135

Clark JE, Ragsdale SW, Ljungdahl LG, Wiegel J (1982) Levels of enzymes involved in the synthesis of acetate from $CO_2$ in *Clostridium thermoautotrophicum*. J Bacteriol 151:507–509

Colby J, Williams E, Turner APF (1985) Applications of CO-utilizing microorganisms. Trends Biotechnol 3:12–17

Cole JR, Fathepure BZ, Tiedje JM (1995) Tetrachloroethene and 3-chlorobenzoate dechlorination activities are co-induced in *Desulfomonile tiedjei* DCB-1. Biodegradation 6:167–172

Connaris H, West SM, Hough DW, Danson MJ (1998) Cloning and expression in *Escherichia coli* of the gene encoding citrate synthase from hyperthermophilic archaeon *Sulfolobus solfataricus*. Extremophiles 2:61–68

Conrad R (1996) Soil microorganisms as controllers of atmospheric trace gases ($H_2$, CO, OCS, $N_2O$ and NO). Microbiol Rev 60:609–640

Conrad R, Seiler W (1980) Role of microorganisms in the consumption and production of atmospheric carbon monoxide by soil. Appl Environ Microbiol 40:437–445

Conrad R, Seiler W, Bunse G, Giehl H (1982) Carbon-monoxide in sea-water (Atlantic-Ocean). J Geophys Res-Oceans Atmos 87:8839–8852

Cook AM, Scholtz R, Leisinger T (1988) Mikrobieller Abbau von halogenierten aliphatischen Verbindungen. GWF Gas Wasserfach: Wasser Abwasser 129:61–69

Cookson JT Jr (1995) Bioremediation engineering: design and application. McGraw-Hill, New York, p 524

Corella J, Orio A, Aznar P (1998) Biomass gasification with air in fluidized bed: reforming of the gas composition with commercial steam reforming catalysts. Ind Eng Chem Res 37:4617–4624

Côté P, Bersillon JL, Huyard A (1989) Bubble-free aeration using membranes: mass transfer analysis. J Membr Sci 47:91–106

Cotter JL, Chinn MS, Grunden AM (2009) Influence of process parameters on growth of *Clostridium Ljungdahlii* and *Clostridium autoetahanogenum* on synthesis gas. Enzyme Microbial Technol. 44:281–288

Cotter JL, Chinn MS, Grunden AM (2009) Ethanol and acetate production by *Clostridium ljungdahlii* and *Clostridium autoethanogenum* using resting cells. Bioprocess Biosyst Eng 32:369–380

Crawford RL, Crawford DL (2005) Bioremediation: principles and applications. Cambridge University Press, Cambridge, p 416

Cui G, Hong W, Zhang J, Li W, Feng Y, Liu Y, Cui Q (2012) Targeted gene engineering in *Clostridium cellulolyticum* H10 without methylation. J Microbial Methods 89:201–208

Dangel W, Schulz H, Diekert G, König H, Fuchs G (1987) Occurrence of corrinoidcontaining membrane proteins in anaerobic bacteria. Arch Microbiol 148:52–56

Daniel SL, Hsu T, Dean SI, Drake HL (1990) Characterization of the hydrogen- and carbon monoxide-dependent chemolithotrophic potentials of the acetogens *Clostridium thermoaceticum* and *Acetogenium kivui*. J Bacteriol 172:4464–4471

Daniell J, Köpke M, Simpson SD (2012) Commercial biomass syngas fermentation. Energies 5:5372–5417

Daniels L, Fuchs G, Thauer RK, Zeikus JG (1977) Carbon monoxide oxidation by methanogenic bacteria. J Bacteriol 132:118–126

Danielsson B, Mosbach K (1987) Theory and application of calorimetric sensors. In: Turner APF, Karube I, Wilson GS (eds) Biosensors: fundamentals and applications. Oxford University Press, London, pp 575–595

Dashekvicz MP, Uffen RL (1979) Identification of a carbon monoxide-metabolizing bacterium as a strain of *Rhodopseudomonas gelatinosa* (Molisch) van Niel. Int J Syst Bacteriol 29:145–148

Datar RP, Shenkman RM, Cateni BG, Huhnke RL, Lewis RS (2004) Fermentation of biomass-generated producer gas to ethanol. Biotechnol Bioeng 86:587–594

Davidova MN, Tarasova NB, Mukhitova FK, Karpilova IU (1994) Carbon monoxide in metabolism of anaerobic bacteria. Can J Microbiol 40:417–425

Davis G, Allen H, Hill O (1983) Bioelectrochemical fuel cell and sensor based on a quinoprotein, alcohol dehydrogenase. Enzyme Microb Technol 5:383–388

Davis J, Vaughan DH, Cardosi MF (1995) Elements of biosensor construction. Enz Microb Tech 17:1030–1035

Degrand C, Miller LL (1980) An electrode modified with polymer bound dopamine which catalyses NADH oxidation. J Am Chem Soc 102:5728–5732

Demler M, Weuster-Botz D (2011) Reaction engineering analysis of hydrogenotrophic production of acetic acid by *Acetobacterium woodii*. Biotechnol Bioeng 108:470–474

Demirbas A (2007) Progress and recent trends in biofuels. Prog Energy Combust Sci 33:1–18

Demirbas MF, Balat M (2006) Recent advances on the production and utilization trends of biofuels: A global perspective. Energy Convers Manage 47:2371–2381

Dennison MJ, Hall JM, Turner APF (1995) Gas-phase microbiosensor for monitoring phenol vapor at ppb levels. Anal Chem 67:3922–3927

Deppenmeier U (2002) Redox-driven proton translocation in methanogenic Archaea. Cell Mol Life Sci 59:1513–1533

Deppenmeier U, Lienard T, Gottschalk G (1999) Novel reactions involved in energy conservation by methanogenic archaea. FEBS Lett 457:291–297

Desai RP, Papoutsakis ET (1999) Antisense RNA strategies for metabolic engineering of *Clostridium acetobutylicum*. Appl Environ Microbial 65:936–945

DeWeerd K, Mandelco AL, Tanner RS, Woese CR, Suflita JM (1990) *Desulfomonile tiedjei* gen. nov. and sp. nov., a novel anaerobic, dehalogenating, sulfate-reducing bacterium. Arch Microbiol 154:23–30

Diekert GB, Thauer R (1978) Carbon monoxide by *Clostridium thermoaceticum* and *Clostridium formiaceticum*. J Bacteriol 136:597–606

Diekert GB, Thayer RK (1982) Carbon monoxide oxidation by *Clostridium thermoaceticum* and *Clostridium formicoaceticum*. J Bacteriol 136:597–606

Divies C (1975) Remarques sur l'oxidation del'ethanol par une 'electrode microbienne' de *Acinobacter xylinium*. Annals Microbiol 126A:175–186

Dobbek H, Gremer L, Kiefersauer R, Huber R, Meyer O (2002) Catalysis at a dinuclear [CuSMo (=O) OH] cluster in a CO dehydrogenase resolved at 1.1-A resolution. Proc Natl Acad Sci USA 99:15971–15976

Donaghy JA, Bronnenmeier K, Soto-Kelly PF, McKay AM (2000) Purification and characterization of an extracellular feruloyl esterase from the thermophilic anaerobe *Clostridium stercorarium*. J Appl Microbiol 88:458–466

Dong H, Tao W, Zhang Y, Li Y (2012) Development of an anhydrotetracycline-inducible gene expression system for solvent-producing *Clostridium acetobutylicum*: a useful tool for strain engineering. Metab Eng 14:59–67

Doukov TI, Iverson TM, Seravalli J, Ragsdale SW, Drennan CL (2002) A Ni-Fe-Cu center in a bifunctional carbon monoxide dehydrogenase/acetyl-CoA synthase. Science 298:567–572

Drake HL (1994) Acetogenesis, acetogenic bacteria, and the acetyl-CoA "Wood/Ljungdahl" pathway: Past and current perspectives. In: Drake HL (ed) Acetogenesis. Chapman and Hall, New York, pp 3–60

Drake HL, Daniel SL (2004) Physiology of the thermophilic acetogen *Moorella thermoacetica*. Res Microbiol 155:422–436

Drake HL, Hu SI, Wood HG (1980) Purification of carbon monoxide dehydrogenase, a nickel enzyme from *Clostridium thermocaceticum*. J Biol Chem 255:7174–7180

Drake HL, Kusel K, Matthies C (2002) Ecological consequences of the phylogenetic and physiological diversities of acetogens. Antonie Van Leeuwenhoek 81:203–213

Drake HL, Küsel K, Matthies C (2006) Acetogenic prokaryotes. In: Dworkin M, Falkow S, Rosenberg E, Schleifer KH, Stackebrandt E (eds) The prokaryotes. Springer, New York, pp 354–420

Drake HL, Gössner AS, Daniel SL (2008) Old acetogens, new light. Ann New York Acad Sci 1125:100–128

Drennan CL, Heo J, Sintchak MD, Schreiter E, Ludden PW (2001) Life on carbon monoxide: X-ray structure of *Rhodospirillum rubrum* Ni-Fe-S carbon monoxide dehydrogenase. Proc Natl Acad Sci USA 98:11973–11978

Drew TB (1981) Advances in chemical engineering. Elsevier, New York, p 451

Du Z, Li H, Gu T (2007) A state of the art review on microbial fuel cells: a promising technology for wastewater treatment and bioenergy. Biotechnol Adv 25:464–482

Dubinin AG, Li F, Li Y, Yu J (1991) A solid state immobilized enzyme polymer membrane microelectrode for measuring lactate ion concentration. Bioelectrochem Bioenerg 25:131–135

Dudynski M, Kwiatkowski K, Bajer K (2012) From feathers to syngas-technologies and devices. Waste Manage 32:685–691

Dűrre P (2007) Biobutanol: an attractive biofuel. Biotechnol J 2:1525–1534

Egli C, Stromeyer S, Cook AM, Leisinger T (1990) Transformation of tetra- and trichloromethane to $CO_2$ by anaerobic bacteria is a non-enzymic process. FEMS Microbiol Lett 68:207–212

Egli C, Tschan T, Scholtz R, Cook AM, Leisinger T (1988) Transformation of tetrachloromethane to dichloromethane and carbon dioxide by *Acetobacterium woodii*. Appl Environ Microbiol 54:2819–2824

Elam CC, Gregoire Padro CE, Sandrock G, Luzzi A, Lindblad P, Hagen EF (2003) Realizing the hydrogen future: the international energy agency's efforts to advance hydrogen energy technologies. Int J Hydrogen Energy 28:601–607

Ellenberger J, Krishna R (2003) Shaken, not stirred, bubble column reactors: enhancement of mass transfer by vibration excitement. Chem Eng Sci 58:705–710

Empadinhas N, da Costa MS (2008) Osmoadaptation mechanisms in prokaryotes: distribution of compatible solutes. Int Microbiol 11:151–161

Faaij APC (2006) Bio-energy in Europe: Changing technology choices. Energy Policy 34:322–342

Faaij A, van Ree R, Waldheim L, Olsson E, Oudhuis A, van Wijk A, Daey-Ouwens C, Turkenburg W (1997) Gasification of biomass wastes and residues for electricity production. Biomass Bioenerg 12:387–407

Fadavi A, Chisti Y (2005) Gas–liquid mass transfer in a novel forced circulation loop reactor. Chem Eng J 112:73–80

Fardeau ML, Salinas MB, L'Haridon S, Jeanthon C, Verhe F, Cayol JL, Patel BK, Garcia JL, Ollivier B (2004) Isolation from oil reservoirs of novel thermophilic anaerobes phylogenetically related to *Thermoanaerobacter subterraneus*: reassignment of *T. subterraneus*, *Thermoanaerobacter yonseiensis*, *Thermoanaerobacter tengcongensis* and *Carboxydibrachium pacificum* to *Caldanaerobacter subterraneus* gen. nov., sp. nov., comb. nov. as four novel subspecies. Int J Syst Evol Microbiol 54:467–474

Fathepure BZ, Boyd SA (1988) Dependence of tetrachloroethylene dechlorination on methanogenic substrate consumption by *Methanosarcina* sp. strain DCM. Appl Environ Microbiol 54:2976–2980

Fathepure BZ, Nengu JP, Boyd SA (1987) Anaerobic bacteria that dechlorinate perchloroethylene. Appl Environ Microbiol 53:2671–2674

Fei Q, Chang HN, Shang L, Choi J, Kim N, Kang J (2011) The effect of volatile fatty acids as a sole carbon source on lipid accumulation by *Cryptococcus albidus* for biodiesel production. Biores Technol 102:2695–2701

Ferenci T, Strøm T, Quayle JR (1975) Oxidation of carbon monoxide by *Pseudomonas methanica*. J Gen Microbiol 91:79–91

Ferry JG (1995) CO Dehydrogenase. Annual Rev Microbiol 49:305–333

Feustel L, Nakotte S, Durre P (2004) Characterization and development of two reporter gene systems for *Clostridium acetobutylicum*. Appl Environ Microbiol 70:798–803

Fischer F, Zillig W, Stetter KO, Schreiber G (1983) Chemolithoautotrophic metabolism of anaerobic extremely thermophilic archaebacteria. Nature 301:511–513

Fischer CR, Klein-Marcuschamer D, Stephanopoulos G (2008) Selection and optimization of microbial hosts for biofuels production. Metabolic Eng 10:295–304

Fockedey E, Wilde JD, Gerin PA (2008) Separation of chemicals produced by acidogenic fermentation. In: The 2008 Annual Meeting. Philadelphia, PA

Fontaine FE, Peterson WH, McCoy E, Johnson MJ, Ritter GJ (1942) A new type of glucose fermentation by *Clostridium thermoaceticum*. J Bacteriol 43:701–715

Fox JD, Kerby RL, Roberts GP, Ludden PW (1996) Characterization of the CO-induced, CO-tolerant hydrogenase from *Rhodospirillum rubrum* and the gene encoding the large subunit of the enzyme. J Bacteriol 178:1515–1524

Fox JD, He Y, Shelver D, Roberts GP, Ludden PW (1996) Characterization of the region encoding the CO-induced hydrogenase *of Rhodospirillum rubrum*. J Bacteriol 178:6200–6208

Franks AE, Nevin K (2010) Microbial fuel cells, a current review. Energies 3:899–919

Freedman DL, Gossett JM (1989) Biological reductive dechlorination of tetrachloroethylene and trichloroethylene to ethylene under methanogenic conditions. Appl Environ Microbiol 55:2144–2151

Frostl JM, Seifritz C, Drake HL (1996) Effect of nitrate on the autotrophic metabolism of the acetogens *Clostridium thermoautotrophicum* and *Clostridium thermoaceticum*. J Bacteriol 178:4597–4603

Frunzke K, Meyer O (1990) Nitrate respiration, denitrification, and utilization of nitrogen sources by aerobic carbon monoxideoxidizing bacteria. Arch Microbiol 154:168–174

Gadkari D, Schricker K, Acker G, Kroppenstedt RN, Meyer O (1990) *Streptomyces thermoautotrophicus* sp. nov., a thermophilic CO- and $H_2$-oxidizing obligate chemolithoautotroph. Appl Environ Microbiol 56:3727–3734

Galli R, McCarty PL (1989) Biotransformation of 1,1,1-trichloroethane, trichloromethane, and tetrachloromethane by a *Clostridium* sp. Appl Environ Microbiol 55:837–844

Gantzer CJ, Wackett LP (1991) Reductive dechlorination catalyzed by bacterial transition-metal coenzymes. Environ Sci Technol 25:715–722

Gavrilescu M, Roman RV, Tudose RZ (1997) Hydrodynamics in external-loop airlift bioreactors with static mixers. Bioprocess Biosyst Eng 16:93–99

Genthner BRS, Bryant MP (1982) Growth of *Eubacterium limosum* with carbon monoxide as the energy source. Appl Environ Microbiol 43:70–74

Genthner BRS, Davis CL, Bryant MP (1981) Features of rumen and sewage sludge strains of *Eubacterium limosum*, a methanol-utilizing and $H_2$-$CO_2$-utilizing species. Appl Environ Microbiol 42:12–19

Genthner BRS, Bryant MP (1987) Additional characteristics of one-carbon compound utilization by *Eubacterium limosum* and *Acetobacterium woodii*. Appl Environ Microbiol 53:471–476

Gentile M, Yan T, Tiquia SM, Fields MW, Nyman J, Zhou J, Criddle CS (2006) Stability in a denitrifying fluidized bed reactor. Microb Ecol 2006 52:311–321

Gilmartin MAT, Hart JP (1995) Development of one-shot biosensors for the measurement of uric acid cholesterol. Anal Proc Anal Commun 32:341–345

Girbal L, Mortier-Barriere I, Raynaud F, Rouanet C, Croux C, Soucaille P (2003) Development of a sensitive gene expression reporter system and an inducible promoter-repressor system for *Clostridium acetobutylicum*. Appl Environ Microbiol 69:4985–4988

Global Industry Analysts Inc. Global market for acetic acid to Reach 12.15 Million Tons by 2017. http://www.prweb.com/releases/acetic_acid_acetates/vinyl_acetate_monomer_PTA/prweb9242731.htm(accessed on 12 July 2014)

Goodson LH, Jacobs WB (1974) Application of immobilized enzymes to detection and monitoring. In: EK Pye, LB Wingard (eds) Enzyme engineering, vol 2. Plenum Press, New York, pp 393–400

Gordillo G, Annamalai K (2010) Adiabatic fixed bed gasification of dairy biomass with air and steam. Fuel 89:384–391

Gorris LGM, van der Drift C, Vogels GD (1988) Separation and quantification of cofactors from methanogenic bacteria by high–performance liquid chromatography: optimum and routine analyses. J Microbiol Methods 8:175–190

Gössner AS, Picardal F, Tanner RS, Drake HL (2008) Carbon metabolism of the moderately acid-tolerant acetogen *Clostridium drakei* isolated from peat. FEMS Microbiol Lett 287:236–242

Grahame DA, DeMoll E (1995) Substrate and accessory protein requirements and thermodynamics of acetyl-CoA synthesis and cleavage in *Methanosarcina barkeri*. Biochemistry 34:4617–4624

Greco C, Bruschi M, Heimdal J, Fantucci P, De Gioia L, Ryde U (2007) Structural insights into the active-ready form of [FeFe]-hydrogenase and mechanistic details of its inhibition by carbon monoxide. Inorg Chem 46:7256–7258

Green EM (2011) Fermentative production of butanol—the industrial perspective. Curr Opin Biotechnol 22:337–343

Greene AC, Patel BKC, Sheehy AJ (1997) *Deferribacter thermophilus* gen. nov., sp. nov., a novel thermophilic manganese- and iron-reducing bacterium isolated from a petroleum reservoir. Int J Syst Bacteriol 47:505–509

Gremer L, Kellner S, Dobbek H, Huber R, Meyer O (2000) Bindingof flavin adenine dinucleotide to molybdenum-containing carbon monoxide dehydrogenase from *Oligotropha carboxidovorans*. J Biol Chem 275:1864–1872

Grethlein A, Worden R, Jain M, Datta R (1990) Continuous production of mixed alcohols and acids from carbon monoxide. Appl Biochem Biotechnol 24–25:875–884

Grethlein AJ, Worden RM, Jain MK, Datta R (1991) Evidence for production of N-butanol from carbon monoxide by *Butyribacterium methylotrophicum*. J Ferment Bioeng 72:58–60

Griffin DW, Schultz MA (2012) Fuel and chemical products from biomass syngas: A comparison of gas fermentation to thermochemical conversion routes. Environ Prog Sustain Ener 31:219–224

Gugliandolo C, Lentini V, Spanò A, Maugeri TL (2012) New bacilli from shallow hydrothermal vents of Panarea Island (Italy) and their biotechnological potential. J Appl Microbiol 112:1102–1112

Guilbault GG (1983) Determination of formaldehyde with an enzyme-coated piezoelectric crystal detector. Anal Chem 55:1682–1684

Guo Y, Xu J, Zhang Y, Xu H, Yuan Z, Li D (2010) Medium optimization for ethanol production with *Clostridium autoethanogenum* with carbon monoxide as sole carbon source. Bioresour Technol 101:8784–8789

Hallas LE, Alexander M (1983) Microbial transformation of nitroaromatic compounds in sewage effluent. Appl Environ Microbiol 45:1234–1241

Hammerle M, Hall EAH (1996) Electrochemical enzyme sensor for formaldehyde operating in the gas phase. Biosens Bioelectron 11:239–246

Hanzelmann P, Dobbek H, Gremer L, Huber R, Meyer O (2000) The effect of intracellular molybdenum in *Hydrogenophaga pseudoflava* on the crystallographic structure of the seleno-molybdo-iron-sulfur flavoenzyme carbon monoxide dehydrogenase. J Mol Biol 301:1221–1235

Haryanto A, Fernando SD, Pordesimo LO, Adhikari S (2009) Upgrading of syngas derived from biomass gasification: A thermodynamic analysis. Biomass Bioener 33:882–889

Hausinger RP (1993) Carbon monoxide dehydrogenase. In: Frieden E (ed) Biochemistry of the elements, vol 12. Springer, New York, pp 107–145

Hayanakawa J, Kondoh Y, Ishizuka M (2009) Cloning and characterization of flagellin genes and identification of flagellin glycosylation from thermophilic *Bacillus* species. Biosci Biotechnol Biochem 73:1450–1452

He Z, Minteer SD, Angenent LT (2005) Electricity generation from artificial wastewater using an upflow microbial fuel cell. Environ Sci Technol 39:5262–5267

Heap JT, Pennington OJ, Cartman ST, Carter GP, Minton NP (2007) The ClosTron: a universal gene knock-out system for the genus *Clostridium*. J Microbiol Method 70:452–464

Heap JT, Kuehne S, Ehsaan M, Cartman ST, Cooksley CM, Scott JC, Minton NP (2010) The ClosTron: mutagenesis in *Clostridium* refined and streamlined. J Microbiol Method 80:49–55

Heap JT, Ehsaan M, Cooksley CM, Ng YK, Cartman ST, Winzer K, Minton NP (2012) Integration of DNA into bacterial chromosomes from plasmids without a counter-selection marker. Nucleic Acid Res 40:1–10

Hedderich R (2004) Energy-converting [NiFe] hydrogenases from archaea andextremophiles: ancestors of complex I. J Bioenerg Biomembr 36:65–75

Hedderich R, Forzi L (2005) Energy-converting [NiFe] hydrogenases: more than just $H_2$ activation. J Mol Microbiol Biotechnol 10:92–104

Heiskanen H, Virkajarvi I, Viikari L (2007) The effects of syngas composition on the growth and product formation of *Butyribacterium methylotrophicum*. Enz Microb Technol 41:362–367

Henstra AM, Stams AJ (2004) Novel physiological features of *Carboxydothermus hydrogenoformans* and *Thermoterrabacterium ferrireducens*. Appl Environ Microbiol 70:7236–7240

Henstra AM, Dijkema C, Stams AJ (2007) *Archaeoglobus fulgidus* couples CO oxidation to sulfate reduction and acetogenesis with transient formate accumulation. Environ Microbiol 9:1836–1841

Henstra AM, Sipma J, Rinzema A, Stams AJ (2007) Microbiology of synthesis gas fermentation for biofuel production. Curr Opin Biotechnol 18:200–206

Heo J, Staples CR, Telser J, Ludden PH (1999) *Rhodospirillum rubrum* CO-dehydrogenase. Part 2. Spectroscopic investigation and assignment of spin-spin coupling signals. J Am Chem Soc 121:11045–11057

Heo J, Staples CR, Telser J, Ludden PH (2000) Evidence for a ligand CO that is required for catalytic activity of CO dehydrogenase from *Rhodospirillum rubrum*. Biochemistry 39:7956–7963

Hemme CL, Mouttaki H, Lee YJ, Zhang G, Goodwin L, Lucas S, Copeland A, Lapidus A, Glavina del Rio T, Tice H, Saunders E, Brettin T, Detter JC, Han CS, Pitluck S, Land ML, Hauser LJ, Kyrpides N, Mikhailova N, He Z, Wu L, Van Nostrand JD, Henrissat B, He Q, Lawson PA, Tanner RS, Lynd LR, Wiegel J, Fields MW, Arkin AP, Schadt CW, Stevenson BS, McInerney MJ, Yang Y, Dong H, Xing D, Ren N, Wang A, Huhnke RL, Mielenz JR, Ding SY, Himmel ME, Taghavi S, van der Lelie D, Rubin EM, Zhou J (2010) Sequencing of multiple *Clostridia* genomes related to biomass conversion and biofuels production. J Bacterial 192:6494–6496

Herrera S (2006) Bonkers about biofuels. Nat Biotechnol 24:755–760

Herrmann I, Kramm UI, Fiechter S, Bogdanoff P (2009) Oxalate supported pyrolysis of CoTMPP as electrocatalysts for the oxygen reduction reaction. Electrochim Acta 54:4275–4287

Hickey R, Datta R, Tsai SP, Basu R (2008) Membrane supported bioreactor for conversion of syngas components to liquid products. US Patent

Hierlemann A, Baltes H (2003) CMOS-based chemical microsensors. The Analyst 128:15–28

Hierlemann A, Brand O, Hagleitner C, Baltes H (2003) Microfabrication techniques for chemical/biosensors. Proc IEEE 91:839–863

Higgins IJ, Best DJ, Hammond RC (1980) New Findings in methane-utilizing bacteria highlight their importance in the biosphere and their commercial potential. Nature 286:561–564

Hikuma M, Kubo T, Yasuda T, Karube I, Suzuki S (1980) Ammonia electrode with immobilized nitrifying bacteria. Anal Chem 52:1020–1024

Hill HA, Walton NJ, Higgins IJ (1981) Electrochemical reduction of dioxygen using a terminal oxidase. FEBS Lett 126:282–284

Horowitz A, Suffita JM, Tiedje JM (1983) Reductive dehalogenations of halobenzoates by anaerobic lake sediment microorganisms. Appl Environ Microbiol 45:1459–1465

Hu P, Bowen SH, Lewis RS (2011) A thermodynamic analysis of electron production during syngas fermentation. Bioresour Technol 102:8071–8076

Hu Z, Spangler NJ, Anderson ME, Xia J, Ludden PW, Lindahl PA, Münckd E (1996) Nature of the C-cluster in Ni-containing carbon monoxide dehydrogenases. J Am Chem Soc 118:830–845

Huang S, Lindahl PA, Wang C, Bennett GN, Rudolph FB, Hughes JB (2000) 2,4,6-trinitrotoluene reduction by carbon monoxide dehydrogenase from *Clostridium thermoaceticum*. Appl Environ Microbiol 66:1474–1478

Huang H, Liu H, Gan YR (2010) Genetic modification of critical enzymes and involved genes in butanol biosynthesis from biomass. Biotechnol Adv 28:651–657

Hubley JH, Mitton JR, Wilkinson JF (1974) The oxidation of carbon monoxide by methane oxidizing bacteria. Arch Microbiol 95:365–368

Huck H, Schmidt HL (1981) Chloranil alskatalysator zur elektrochemischen oxidation von NADH zu NAD$^+$. Angewandt Chemie 93:421–422

Hugendieck I, Meyer O (1992) The structural genes encoding CO dehydrogenase subunits (cox L, M and S) in *Pseudomonas carboxydovorans* OM5 reside on plasmid phCG3 and are, with the exception of *Streptomyces thermoautotrophicus*, conserved in carboxydotrophic bacteria. Arch Microbiol 157:301–304

Huhnke R, Lewis RS, Tanner RS (2008) Isolation and characterization of novel clostridial species. US Patent

Hurst KM, Lewis RS (2010) Carbon monoxide partial pressure effects on the metabolic process of syngas fermentation. Biochem Eng J 48:159–165

Hussain A, Guiot SR, Mehta P, Raghavan V, Tartakovsky B (2011) Electricity generation from carbon monoxide and syngas in a microbial fuel cell. Appl Microbiol Biotechnol 90:827–836

Ide A, Niki Y, Sakamoto F, Watanabe I, Watanabe H (1972) Decomposition of pentachlorophenol in paddy soil. Agric Biol Chem 36:1937–1944

Ieropoulos I, Winfield J, Greenman J (2010) Effects of flow-rate, inoculum and time on the internal resistance of microbial fuel cells. Bioresour Technol 101:3520–3525

Imhoff JF, Trüper HG, Pfennig N (1984) Rearrangement of the species and genera of the phototrophic purple nonsulfur bacteria. Int J Syst Bacteriol 34:340–343

Imkamp F, Biegel E, Jayamani E, Buckel W, Müller V (2007) Dissection of the caffeate respiratory chain in the acetogen Acetobacterium woodii: identification of an Rnf-type NADH dehydrogenase as a potential coupling site. J Bacterial 189:8145–8153

Inman RE, Ingersoll RB (1971) Uptake of carbon monoxide by soil fungi. J Air Pollut Control Assoc 21:646–647

Inokuma K, Nakashimada Y, Akahoshi T, Nishio N (2007) Characterization of enzymes involved in the ethanol production of Moorella sp. HUC22-1. Arch Microbiol 188:37–45

Ismail KSK, Najafpour G, Younesi H, Mohamed AR, Kamaruddin AH (2008) Biological hydrogen production from CO: Bioreactor performance. Biochem Eng J 39:468–477

IPCC (2001) Climate Change 2001. The scientific basis. contribution of working group 1 to the third assessment report of the intergovernmental panel of climate change. Cambridge University Press, New York, p 398

Jablonski PE, Ferry JG (1992) Reductive dechlorination of trichloroethylene by the CO-reduced CO dehydrogenase enzyme complex from *Methanosarcina thermophila*. FEMS Microbiol Lett 96:55–60

Jablonski PE, Ferry JG (1993) Use of carbon monoxide dehydrogenase for bioremediation of toxic compounds US Patent 5466600 A

Jaegfeldt H, Torstensson A, Gorton L, Johansson G (1981) Catalytic oxidation of reduced nicotinamide adenine dinucleotide (NADH) by graphite electrodes modified with adsorbed aromatics containing catechol functionalities. Anal Chem 53:1979–1982

Jaegfeldt H, Wuwana T, Johansson G (1983) Electrochemical stability of catechols with pyrene side chain in strongly adsorbed on graphite electrodes for catalytic oxidation of dihydronicotinamide adenine dinucleotide. J Am Chem Soc 105:1805–1814

Jang YS, Lee J, Malaviya A, Seung do Y, Cho JH, Lee SY. (2012) Butanol production from renewable biomass: Rediscovery of metabolic pathways and metabolic engineering. Biotechnol J 7:186–198

Jiang B, Henstra AM, Paulo PL, Balk M, van Doesburg W, Stams AJM (2009) A typical one-carbon metabolism of an acetogenic and hydrogenogenic *Moorella thermoacetica* strain. Arch Microbiol 191:123–131

Jin G, Yang F, Hu C, Shen H, Zhao ZK (2012) Enzyme-assisted extraction of lipids directly from the culture of the oleaginous yeast *Rhodosporidium toruloides*. Bioresour Technol 111:378–382

Jones DT, Woods DR (1986) Acetone-butanol fermentation revisited. Microbiolog Rev 50:484–524

Jones RD, Morita RY (1983) Carbon monoxide oxidation by chemolithotrophic ammonium oxidizers. Can J Microbiol 29:1545–1551

Jong BC, Kim BH, Chang IS, Liew PWY, Choo YF, Kang GS (2006) Enrichment, performance, and microbial diversity of a thermophilic mediatorless microbial fuel cell. Environ Sci Technol 40:6449–6454

Jung GY, Jung HO, Kim JR, Ahn Y, Park S (1999) Isolation and characterization of *Rhodopseudomonas palustris* P4 which utilizes CO with the production of $H_2$. Biotechnol Lett 21:525–529

Jung GY, Kim JR, Jung HO, Park JY, Park S (1999) A new chemoheterotrophic bacterium catalyzing water-gas shift reaction. Biotechnol Lett 21:869–873

Jung GY, Kim JR, Park JY, Park S (2002) Hydrogen production by a new chemoheterotrophic bacterium *Citrobacter* sp. Y19. Int J Hydrogen Energy 27:601–610

Kaisheva A, Iliev I, Christov S, Kazareva R (1997) Electrochemical gas biosensor for phenol. Sensor Actuators 44:571–577

Kaneko T, Nakamura Y, Sato S, Minamisawa K, Uchiumi T, Sasamoto S, Watanabe A, Idesawa K, Iriguchi M, Kawashima K, Kohara M, Matsumoto M, Shimpo S, Tsuruoka H, Wada T, Yamada M, Tabata S (2002) Complete genomic sequence of nitrogen-fixing symbiotic bacterium *Bradyrhizobium japonicum* USDA110. DNA Res 9:189–197

Kane MD, Breznak JA (1991) *Acetonema longum* gen nov SP-nov, an $H_2/CO_2$ acetogenic bacterium from the termite, *Pterotermes occidentis*. Arc Microbiol 156:91–98

Kang BS, Kim YM (1999) Cloning and molecular characterization of the genes for carbon monoxide dehydrogenase and localization of molybdopterin, flavin adenine dinucleotide, and iron–sulfur centers in the enzyme of *Hydrogenophaga pseudoflava*. J Bacteriol 181:5581–5590

Kaplan DL, Kaplan AM (1982) Thermophilic biotransformatlons of 2,4,6-trinitrotoluene under simulated composting conditions. Appl Environ Microbiol 44:757–760

Kashefi K, Lovley DR (2000) Reduction of Fe(III), Mn(IV), and toxic metals at 100 °C by Pyrobaculum islandicum. Appl Environ Microbiol 66:1050–1056

Karube I, Okada T, Suzuki S (1982) A methane gas sensor based on oxidizing bacteria Anal Chim Acta 135:61–67

Karyakin AA, Karyakina EE, Schuhmann W, Schmidt HL, Varfolomeyev SD (1994) New amperomtric dehydrogenase electrodes based on electrocatalytic NADH-oxidation at poly(-methylene blue)-modified electrodes. Electroanal 6:821–829

Karyakin AA, Gitelmacher OV, Karyakina EE (1995) Prussian blue based first generation biosensors, A high sensitive amperometric electrode for glucose. Anal Chem 67:2419–2423

Kashefi K, Holmes DE, Baross JA, Lovley DR (2003) Thermophily in the *Geobacteraceae*: *Geothermobacter ehrlichii* gen. nov., sp. nov., a novel thermophilic member of the *Geobacteraceae* from the "Bag City" hydrothermal vent. Appl Environ Microbiol 69:2985–2993

Kazy SK, Das SK, Sar P (2006) Lanthanum biosorption by a *Pseudomonas* sp.: equilibrium studies and chemical characterization. J Ind Microbiol Biotechnol 33:773–783

Kerby R, Zeikus JG (1983) Growth of *Clostridium thermoaceticum* on $H_2/CO_2$ or CO as energy source. Curr Microbiol 8:27–30

Kerby R, Zeikus JG (1987) Catabolic enzymes of the acetogen *Butyribacterium methylotrophicum* grown on single-carbon substrates. J Bacteriol 169:5605–5609

Kerby RL, Ludden PW, Roberts GP (1995) Carbon monoxidedependent growth of *Rhodospirillum rubrum*. J Bacteriol 177:2241–2244

Kim YM, Hegeman GD (1981) Purification and some properties of carbon monoxide dehydrogenase from Pseudomonas carboxydohydrogena. J Bacteriol 148:904–911

Kim YM, Hegeman GD (1983) Oxidation of carbon monoxide by bacteria. Int Rev Cytol 81:1–32

Kim D, Chang I (2009) Electricity generation from synthesis gas by microbial processes: CO fermentation and microbial fuel cell technology. Bioresour Technol 100:4527–4530

Kim YM, Park SW (2012) Microbiology and genetics of CO utilization in mycobacteria. Antonie Van Leeuwenhoek 101:685–700

King GM (1999) Attributes of atmospheric carbon monoxide oxidation by Maine forest soils. Appl Environm Microbiol 65:5257–5264

King GM (2003) Uptake of carbon monoxide and hydrogen at environmentally relevant concentrations by mycobacteria. Appl Environ Microbiol 69:7266–7272

King GM, Weber CF (2007) Distribution, diversity and ecology of aerobic CO-oxidizing bacteria. Nat Rev Microbiol 5:107–118

Kirkpatrick D, Biggs SR, Conway B, Finn CM, Hawkins DR, Honda T, Ishida M, Powell GP (1981) Metabolism of N-(2,3-dichlorophenyl)-3,4,5,6-tetrachlorophthalamic acid (Techlofthalam) in paddy soil and rice. J Agric Food Chem 29:1149–1153

Klasson KT, Ackerson CMD, Clausen EC, Gaddy JL (1990) Bioreactor design for synthesis gas fermentation. Biotechnology for the production of clean fuels, Washington, pp 885–900

Klasson KT, Ackerson MD, Clausen EC, Gaddy JL (1991) Bioreactors from synthesis gas fermentations. Resour Conserv Recycling 5:145–165

Klasson KT, Ackerson MD, Clausen EC, Gaddy JL (1992) Bioconversion of synthesis gas into liquid or gasesous fuels. Enz Micr Technol 14:602–608

Klasson KT, Ackerson MD, Clausen EC, Gaddy JL (1993) Biological conversion of coal and coal-derived synthesis gas. Fuel 72:1673–1678

Klemps R, Cypionka H, Widdel F, Pfennig N (1985) Growth with hydrogen, and further physiological characteristics of *Desulfotomaculum* species. Arch Microbiol 143:203–208

Klenk HP, Clayton RA, Tomb JF, White O, Nelson KE, Ketchum KA, Dodson RJ, Gwinn M, Hickey EK, Peterson JD, Richardson DL, Kerlavage AR, Graham DE, Kyrpides NC, Fleischmann RD, Quackenbush J, Lee NH, Sutton GG, Gill S, Kirkness EF, Dougherty BA, McKenney K, Adams MD, Loftus B, Peterson S, Reich CI, McNeil LK, Badger JH, Glodek A, Zhou L, Overbeek R, Gocayne JD, Weidman JF, McDonald L, Utterback T, Cotton MD, Spriggs T, Artiach P, Kaine BP, Sykes SM, Sadow PW, D'Andrea KP, Bowman C, Fujii C, Garland SA, Mason TM, Olsen GJ, Fraser CM, Smith HO, Woese CR, Venter JC (2007) The complete genome sequence of the hyperthermophilic, sulphate-reducing archaeon *Archaeoglobus fulgidus*. Nature 390:364–370

Kobayashi H, Endo K, Sakata S, Mayumi D, Kawaguchi H, Ikarashi M, Miyagawa Y, Maeda H, Sato K (2011) Phylogenetic diversity of microbial communities associated with the crude-oil, large-insoluble-particle and formation-water components of the reservoir fluid from a non-flooded high-temperature petroleum reservoir. J Biosci Bioeng 113:204–210

Kochetkova TV, Rusanov II, Pimenov NV, Kolganova TV, Lebedinsky AV, Bonch-Osmolovskaya EA, Sokolova TG (2011) Anaerobic transformation of carbon monoxide by microbial communities of Kamchatka hot springs. Extremophiles 15:319–325

Koncki R, Radomska A, Glab S (2000) Potentiometric determination of dialysate urea nitrogen. Talanta 52:13–17

Kongjan P, O-Thong S, Kotay M, Min B, Angelidaki I (2010) Biohydrogen production from wheat straw hydrolysate by dark fermentation using extreme thermophilic mixed culture. Biotechnol Bioeng 105:899–908

Köpke M, Liew F (2012) Production of Butanol from Carbon Monoxide by a RecombinantMicroorganism. Patent WO/2012/053905

Köpke M, Held C, Hujer S, Liesegang H, Wiezer A, Wollherr A, Ehrenreich A, Liebl W, Gottschalk G, Dürre P (2010) *Clostridium ljungdahlii* represents a microbial production platform based on syngas. Proc Natl Acad Sci USA 107:13087–13092

Köpke M, Noack S, Dürre P (2011) The past, present, and future of biofuels—Biobutanol as promising alternative. In: dos Santos Bernades MA (ed) Biofuel production-Recent developments and prospects. InTech Rijeka, Croatia, pp 451–486

Koskinen PE, Lay CH, Puhakka JA, Lin PJ, Wu SY, Orlygsson J, Lin CY (2008) High-efficiency hydrogen production by an anaerobic, thermophilic enrichment culture from an Icelandic hot spring. Biotechnol Bioeng 101:665–678

Kotay SM, Das D (2008) Biohydrogen as a renewable energy resource—Prospects and potentials. Int J Hydro Energy 33:258–263

Knoll G, Winter J (1988) Anaerobic degradation of phenol in sewage sludge. Benzoate formation from phenol and CO, in the presence of hydrogen. Appl Microbiol Biotechnol 25:384–391

Krichnavaruk S, Pavasant P (2002) Analysis of gas–liquid mass transfer in an airlift contactor with perforated plates. Chem Eng J 89:203–211

Krone UE, Thauer RK (1992) Dehalogenation of trichlorofluoromethane (CFC-11) by Methanosarcina barkeri. FEMS Microbiol Lett 90:201–204

Krone UE, Thauer RK, Hogenkamp HPC (1989) Reductive dehalogenation of chlorinated C1-hydrocarbons mediated by corrinoids. Biochemistry 28:4908–4914

Krone UE, Thauer RK, Hogenkamp HPC, Steinbach K (1991) Reductive formation of carbon monoxide from $CCl_4$ and FREONs 11, 12 and 13 catalyzed by corrinoids. Biochemistry 30:2713–2719

Krueger B, Meyer O (1984) Thermophilic bacilli growing with carbon monoxide. Arch Microbiol 139:402–408

Krumholz LR, Bryant MP (1985) Clostridium pfennigii sp. nov. uses methoxyl groups of monobenzenoids and produces butyrate. Int J Syst Bact 35:454–456

Krzycki JA, Wolkin RH, Zeikus JG (1982) Comparison of unitrophic and mixotrophic substrate metabolism by acetate-adapted strain of Methanosarcina barkeri. J Bacteriol 149:247–254

Kuehne SA, Heap JT, Cooksley CM, Cartman ST, Minton NP (2011) ClosTron-mediated engineering of Clostridium. Methods Mol Biol 765:389–407

Kumar A (2000) Biosensors based on piezoelectric crystal detectors: theory and application. JOMe, 52 (10): 1–6

Kundiyana DK, Huhnke RL, Wilkins MR (2010) Syngas fermentation in a 100-L pilot scale fermentor: design and process considerations. J Biosci Bioeng 109:492–498

Kundiyana DK, Huhnke RL, Wilkins MR (2011) Effect of nutrient limitation and two-stage continuous fermentor design on productivities during Clostridium ragsdalei syngas fermentation. Bioresour Technol 102:6058–6064

Kundiyana DK, Wilkins MR, Maddipati P, Huhnke RL (2011) Effect of temperature, pH and buffer presence on ethanol production from synthesis gas by Clostridium ragsdalei. Bioresour Technol 102:5794–5799

Küsel K, Dorsch T, Acker G, Stackebrandt E, Drake HL (2011) Clostridium scatologenes strain SL1 isolated as an acetogenic bacterium from acidic sediments. Int J Syst Evol Microbiol 2:537–546

Lapidus AL, Grobovenko SY, Mukhitova FK, Kiyashko SV (1989) Enzyme-catalyzed synthesis of hydrocarbons and oxygen-containing compounds from CO and $H_2$. J Molec Catal 56:260–265

Lederle SM (2010) Heterofermentative Acetonproduktion. Ph.D. Thesis, Ulm University, Ulm, Germany

Lee JI, Karube I (1996) Development of a biosensor for gaseous cyanide in solution. Biosens Bioelectron 11:1147–1154

Lee HS, Kang SG, Bae SS, Lim JK, Cho Y, Kim YJ, Jeon JH, Cha SS, Kwon KK, Kim HT, Park CJ, Lee HW, Kim SI, Chun J, Colwell RR, Kim SJ, Lee JH (2008) The complete genome sequence of Thermococcus onnurineus NA1 reveals a mixed heterotrophic and carboxydotrophic metabolism. J Bacteriol 190:7491–7499

Leibold H, Hornung A, Seifert H (2008) HTHP syngas cleaning concept of two stage biomass gasification for FT synthesis. Powder Technol 180:265–270

Leigh JA, Wolfe RS (1983) Acetogenium kivui gen. nov., sp. nov., a Thermophilic Acetogenic Bacterium. Int J Syst Bacteriol 33:886

Leigh JA, Mayer F, Wolfe RS (1981) Acetogenium kivui, a new thermophilic hydrogen-oxidizing acetogenic bacterium. Arch Microbiol 129:275–280

Lessner DJ, Li L, Li Q, Rejtar T, Andreev VP, Reichlen M, Hill K, Moran JJ, Karger BL, Ferry JG (2006) An unconventional pathway for reduction of $CO_2$ to methane in CO-grown Methanosarcina acetivorans revealed by proteomics. Proc Natl Acad Sci USA 103:17921–17926

Levy PF, Barnard GW, Garcia-Martinez DV, Sanderson JE, Wise DL (1981) Organic acid production from $CO_2/H_2$ and $CO/H_2$ by mixed-culture anaerobes. Biotechnol Bioeng 23:2293–2306

Li WF, Zhou XX, Lu P (2005) Structural features of thermozymes. Biotechnol Adv 23:271–281

Liang Y, Feng Z, Yesuf J, Blackburn JW (2010) Optimization of growth medium and enzyme assay conditions for crude cellulases produced by a novel thermophilic and cellulolytic bacterium, *Anoxybacillus* sp. 527. Appl Biochem Biotechnol 160:1841–1852

Liao BQ, Xie K, Lin HJ, Bertoldo D (2010) Treatment of kraft evaporator condensate using a thermophilic submerged anaerobic membrane bioreactor. Water Sci Technol 61:2177–2183

Liew FM, M Köpke, SD Simpson (2013). Gas Fermentation for Commercial Biofuels Production, Liquid, Gaseous and Solid Biofuels-Conversion Techniques, Prof. Zhen Fang (Ed.), ISBN: 978-953-51-1050-7, InTech, DOI: 10.5772/52164. Available from: http://www.intechopen. com/ books/ liquid-gaseous-and-solid-biofuels-conversion-techniques/gas-fermentation-for-commercial-biofuels-production

Lindahl PA (2002) The Ni-containing carbon monoxide dehydrogenase family: light at the end of the tunnel. Biochem 41:2097–2105

Lindahl PA, Chang B (2001) The evolution of acetyl-CoA synthase. Orig Life Evol Biosph 31:403–434

Lindgren A, Ruzgas T, Gorton L, Csoregi E, Ardila GB, Sakharov IY, Gazaryan IG (2000) Biosensors based on novel peroxidases with improved properties in direct and mediated electron transfer. Biosens Bioelectron 15:491–497

Liou JS, Balkwill DL, Drake GR, Tanner RS (2005) *Clostridium carboxidivorans* sp. nov., a solvent-producing clostridium isolated from an agricultural settling lagoon, and reclassification of the acetogen *Clostridium scatologenes* strain SL1 as *Clostridium drakei* sp. nov. Int J Syst Evol Microbiol 55:2085–2091

Littlechild JA (2011) Thermophilic archaeal enzymes and applications in biocatalysis. Biochem Soc Trans 39:155–158

Liu H, Logan BE (2004) Electricity generation using an air-cathode single chamber microbial fuel cell in the presence and absence of a proton exchange membrane. Environ Sci Technol 38:4040–4046

Liu K, Atiyeh HK, Tanner RS, Wilkins MR, Huhnke RL (2012) Fermentative production of ethanol from syngas using novel moderately alkaliphilic strains of *Alkalibaculum bacchi*. Biores Technol 104:336–341

Ljungdahl LG (1986) The autotrophic pathway of acetate synthesis in acetogenic bacteria. Ann Rev Microbial 40:415–450

Ljungdahl LG (1994) The acetyl-CoA pathway and the chemiosmotic generation of ATP during acetogenesis. In: Ferry JG (ed) acetogenesis. Chapman & Hall, New York, pp 63–87

Logan B (2008) Microbial Fuel Cells, Wiley, Hoboken, New Jersey, p 216

Logan BE (2009) Exoelectrogenic bacteria that power microbial fuel cells. Nat Rev Micro 7:375–381

Logan BE, Regan JM (2006) Electricity-producing bacterial communities in microbial fuel cells. Trends Microbiol 14:512–518

Lorenzo E, Pariente F, Hernandez L, Tobalina F, Darder M, Wu Q, Maskus M, Abruna HD (1998) Analytical starategies for amperometric biosensors based on chemically modified electrodes. Biosens Bioelectron 13:319–332

Lorite MJ, Tachil J, Sanjuan J, Meyer O, Bedmar EJ (2000) Carbon monoxide dehydrogenase activity in *Bradyrhizobium japonicum*. Appl Environ Microbiol 68:1871–1876

Lorowitz WH, Bryant MP (1984) *Peptostreptococcus productus* strain that grows rapidly with CO as the energy source. Appl Environ Microbiol 47:961–964

Lovley DR (2006) Bug juice: harvesting electricity with microorganisms. Nat Rev Micro 4:497–508

Lovley DR (2008) The microbe electric: conversion of organic matter to electricity. Curr Opin Biotechnol 19:564–571

Lovley DR, Holmes DE, Nevin KP (2004) Dissimilatory Fe(III) and Mn(IV) Reduction. Adv Microb Physiol 49:219–286

Lowe SE, Jain MK, Zeikus JG (1993) Biology, ecology and biotechnological applications of anaerobic bacteria adapted to environmental stresses in temperature, pH, salinity or substrates. Microbiol Rev 57:451–509

Lupton FS, Conrad R, Zeikus JG (1984) CO metabolism of *Desulfovibrio vulgaris* strain Madison: physiological function in the absence or presence of exogenous substrates. FEMS Microbiol Lett 23:263–268

Lundie LL Jr, Drake HL (1984) Development of a minimally defined medium for the acetogen *Clostridium thermoaceticum*. J Bacteriol 159:700–703

Lütke-Eversloh T, Bahl H (2011) Metabolic engineering of *Clostridium acetobutylicum*: Recent advances to improve butanol production. Curr Opin Biotechnol 22:634–647

Lux MF, Drake HL (1992) Reexamination of the metabolic potentials of the acetogens *Clostridium aceticum* and *Clostridium formicoaceticum* - chemolithoautotrophic and aromatic- dependent growth. FEMS Microbiol Lett 95:49–56

Lynd LR (2008) Energy biotechnology: Editorial overview. Curr Opin Biotechnol 19:199–201

Lynd L, Kerby R, Zeikus JG (1982) Carbon monoxide metabolismof the methylotrophic acidogen, *Butyribacterium methylotrophicum*. J Bacteriol 149:255–263

Lyon EJ, Shima S, Buurman G, Chowdhuri S, Batschauer A, Steinbach K, Thauer RK (2004) UV-A/blue-light inactivation of the 'metal-free' hydrogenase (Hmd) from methanogenic archaea - The enzyme contains functional iron after all. Eur J Biochem 271:195–204

Lyons CM, Colby JP, Williams E (1984) Isolation and characterization and autotrophic metabolism of a moderately thermophilic caboxydobacterium, *Pseudomonas thermo-carboxydovorans* sp. nov. J Gen Microbiol 130:1097–1105

Malin C, Illmer P (2008) Ability of DNA content and DGGE analysis to reflect the performance condition of an anaerobic biowaste fermenter. Microbiol Res 163:503–511

Maness PC, Weaver P (2002) Hydrogen production from a carbon-monoxide oxidation pathway in *Rubrivivax gelatinosus*. Int J Hydrogen Energy 27:1407–1411

Maness PC, Huang J, Smolinski S, Tek V, Vanzin G (2005) Energy Generation from the CO Oxidation-Hydrogen Production Pathway in Rubrivivax gelatinosus. Appl Environ Microbiol 71:2870–2874

Margesin R, Schinner F (2001) Bioremediation (natural attenuation and biostimulation) of diesel-oil-contaminated soil in an alpine glacier skiing area. Appl Environ Microbiol 67:3127–3133

Markov S, Weaver P (2008) Bioreactors for $H_2$ production by purple nonsulfur bacteria. Appl Biochem Biotechnol 145:79–86

Marshall CW, May HD (2009) Electrochemical evidence of direct electrode reduction by a thermophilic Gram-positive bacterium, *Thermincola ferriacetica*. Energy Environ Sci 2:699–705

Marsili E, Baron DB, Shikhare ID, Coursolle D, Gralnick JA, Bond DR (2008) *Shewanella* secretes flavins that mediate extracellular electron transfer. Proc Natl Acad Sci USA 105:3968–3973

Mascini M (1995) Potentiometry: enzyme electrodes. In: Townshend A (ed) Encyclopedia of analytical Science. Academic Press, UK, pp 4112–4118

Mathis B, Marshall C, Milliken C, Makkar R, Creager S, May H (2008) Electricity generation by thermophilic microorganisms from marine sediment. Appl Microbiol Biotechnol 78:147–155

Matsumoto T, Kabe R, Nonaka K, Ando T, Yoon KS, Nakai H, Ogo S (2011) Model study of CO inhibition of [NiFe] hydrogenase. Inorg Chem 50:8902–8906

Mazumder TK, Nishio N, Nagai S (1985) Carbon monoxide conversion to formate by *Methanosarcina barkeri*. Biotechnol Lett 7:377–382

McCormick NG, Feeherry FE, Levmson HS (1976) Microbial transformation of 2,4,6-trinitrotoluene and other nitroaromatic compounds. Appl Environ Microbiol 31:949–958

McKendry P (2002) Energy production from biomass (Part 3): Gasification technologies. Biores Technol 83:55–63

McLafferty FW (1993) Interpretation of mass spectra. University Science Books, California 371 p

Mechichi T, Labat M, Patel BKC, Woo THS, Thomas P, Garcia JL (1999) *Clostridium methoxybenzovorans* sp nov., a new aromatic o-demethylating homoacetogen from an olive mill wastewater treatment digester. Int J Syst Bacteriol 49:1201–1209

Megharaj M, Ramakrishnan B, Venkateswarlu K, Sethunathan N, Naidu R (2011) Bioremediation approaches for organic pollutants: a critical perspective. Environ Int 37:1362–1375

Mehta P, Hussain A, Raghavan V, Neburchilov V, Wang H, Tartakovsky B, Guiot S (2010) Electricity generation from a carbon monoxide in a single chamber microbial fuel cell. Enzyme Microb Technol 46:450–455

Meyer O (1989) Aerobic carbon monoxide-oxidizing bacteria. In: Schlegel HG, Bowien B (eds) Autotrophic bacteria. Science Tech Publishers, Madison, Wisconsin, pp 331–350

Meyer O, Schlegel HG (1980) Carbon monoxide:methylene blue oxidoreductase from *Pseudomonas carboxydovorans*. J Bacteriol 141:74–80

Meyer O, Schlegel HG (1983) Biology of aerobic carbon monoxide-oxidizing bacteria. Annu Rev Microbiol 37:277–310

Meyer O, Frunzke K, Gadkari D, Jacobitz S, Hugendieck I, Kraut M (1990) Utilization of carbon monoxide by aerobes – recent advances. FEMS Microbiol Rev 87:253–260

Min B, Román Ó, Angelidaki I (2008) Importance of temperature and anodic medium composition on microbial fuel cell (MFC) performance. Biotechnol Lett 30:1213–1218

Minunni M, Skladal P, Mascini M (1994) A piezoelectric quartz crystal biosensor as a direct affinity sensor. Anal Lett 27:1475–1487

Mochimaru H, Yoshioka H, Tamaki H, Nakamura K, Kaneko N, Sakata S, Imachi H, Sekiguchi Y, Uchiyama H, Kamagata Y (2007) Microbial diversity and methanogenic potential in a high temperature natural gas field in Japan. Extremophiles 11:453–461

Mohammadi M, Najafpour GD, Younesi F, Lahijani P, Uzir MH, Mohamed AR (2011) Bioconversion of synthesis gas to second generation biofuels: A review. Renewable Sustain Ener Rev 15:4255–4273

Molina CR, Boujtita M, Murr NE (1999) A carbon paste electrode modified by entrapped toluidine blue O for amperometric determination of L-lactate. Anal Chim Acta 401:155–162

Moon H, Chang IS, Kim BH (2006) Continuous electricity production from artificial wastewater using a mediator-less microbial fuel cell. Biores Technol 97:621–627

Moran JJ, House CH, Vrentas JM, Freeman KH (2008) Methyl sulfide production by a novel carbon monoxide metabolism in *Methanosarcina acetivorans*. Appl Environ Microbiol 74:540–542

Moreira AR, Ulmer DC, Linden JC (1981) Butanol toxicity in the butylic fermentation. Biotechnol Bioeng Sym 11:567–579

Muller R, Lingens F (1986) Mikrobieller Abbau halogenierter Kohlenwasserstoffe: ein Beitrag zur Losung vieler Umweltprobleme? Angew Chem 98:778–787

Müller V, Imkamp F, Biegel E, Schmidt S, Dilling S (2008) Discovery of a ferredoxin:NAD+-oxidoreductase (Rnf) in *Acetobacterium woodii*: a novel potential coupling site in acetogens. Ann. New York Acad Sci 1125:137–146

Munasinghe PC, Khanal SK (2010) Biomass-derived syngas fermentation into biofuels: Opportunities and challenges. Bioresour Technol 101:5013–5022

Murthy NBK, Kaufman DD, Fries GF (1979) Degradation of pentachlorophenol (PCP) in aerobic and anaerobic soil. J Environ Sci Health Part B 14:1–14

Naessens M, Minh CT (1998) Whole-cell biosensor for direct determination of solvent vapours. Biosens Bioelectron 13:341–346

Najafpour G, Younesi H (2006) Ethanol and acetate synthesis from waste gases using batch culture of *Clostridium ljungdhahlii*. Enzyme Microb Technol 38:223–228

Najafpour G, Younesi H (2007) Bioconversion of synthesis gas to hydrogen using a light-dependent photosynthetic bacterium, *Rhodospirillum rubrum*. World J Microbiol Biotechnol 23:275–284

Najafpour G, Younesi H, Mohamed AR (2003) Continuous hydrogen production via fermentation of synthesis gas. Petroleum Coal 45:154–158

Najafpour G, Younesi H, Mohamed AR (2004) Effect of organic substrate on hydrogen production from synthesis gas using *Rhodospirillum rubrum*, in batch culture. Biochem Eng J 21:123–130

Nakaminami T, Kuwabata S, Yoneyama H (1997) Electrochemical oxidation of cholesterol catalyzed by cholesterol oxidase with use of an artificial electron mediator. Anal Chem 69:2367–2372

Nath K, Das D (2004) Improvement of fermentative hydrogen production: various approaches. Appl Microbiol Biotechnol 65:520–529

Naumova RP, Selivanovskaya SY, Cherepneva IE (1989) Conversion of 2,4,6-trinitrotoluene under conditions of oxygen and nitrate respiration of Pseudomonas fluorescens. Appl Biochem Microbiol 24:409–413

Nguyen S, Ala F, Cardwell C, Cai D, McKindles KM, Lotvola A, Hodges S, Deng Y, Tiquia-Arashiro SM (2013) Isolation and screening of carboxydotrophs isolated from composts and their potential for butanol synthesis. Environ Technol 34:1995–2007

Nölling J, Hahn D, Ludwig W, De Vos WM (1993) Phylogenetic analysis of thermophilic *Methanobacterium* sp.: Evidence for a formate-utilizing ancestor. Syst Appl Microbiol 16:208–215

Novikov AA, Sokolova TG, Lebedinsky AV, Kolganova TV, Bonch-Osmolovskaya EA (2011) *Carboxydothermus islandicus* sp. nov., a thermophilic, hydrogenogenic, carboxydotrophic bacterium isolated from a hot spring. Int J Syst Evol Microbiol 61:2532–2537

Occupational Safety and Health Standards (1997) 29 CFR Part 1910.1000, Limits for Air Contaminants, Table Z-1

O'Brien RW, Morris JG (1971) The Ferredoxin-dependent reduction of chloramphenicol by *Clostridium acetobutylicum*. J Gen Microbiol 67:265–271

O'Brien JM, Wolkin RH, Moench TT, Morgan JB, Zeikus JG (1984) Association of hydrogen metabolism with unitrophic or mixotrophic growth of *Methanosarcina barkeri* on carbon monoxide. J Bacteriol 158:373–375

Oelgeschlager E, Rother M (2008) Carbon monoxide-dependent energy metabolism in anaerobic bacteria and archaea. Arch Microbiol 190:257–269

Ohwaki K, Hungate RE (1977) Hydrogen utilization by *Clostridia* in sewage sludge. Appl Environ Microbiol 33:1270–1274

Okada T, Karube I, Suzuki S (1981) Microbial sensor system which uses Methylomonas sp. for the determination of methane. Eur J Appl Microbiol Biotechnol 12:102–106

Olson DG, McBride JE, Shaw AJ, Lynd LR (2012) Recent progress in consolidated bioprocessing. Curr Opin Biotechnol 23:396–405

Ormerod MR (2003) Solid oxide fuel cells. Chem Soc Rev 32:17–28

Pakpour F, Najafpour G, Tabatabaei M, Tohidfar M, Younesi H (2014) Biohydrogen production from CO-rich syngas via a locally isolated *Rhodopseudomonas palustris* PT. Bioprocess Biosyst Eng 37:923–930

Pant D, Van Bogaert G, Diels L, Vanbroekhoven K (2010) A review of the substrates used in microbial fuel cells (MFCs) for sustainable energy production. Bioresour Technol 101:1533–1543

Papastathopoulos DS, Rechnitz GA (1975) Enzymatic cholesterol determination using ion selective membrane electrodes. Anal Chem 47:1792–1796

Park SW, Hwang EH, Park H, Kim JA, Heo J, Lee KH, Song T, Kim E, Ro YT, Kim SW, Kim YM (2003) Growth of mycobacteria on carbon monoxide and methanol. J Bacteriol 185:142–147

Park JK, Yee HJ, Kim ST (1995) Amperometric biosensor for determination of ethanol vapor. Biosens Bioelectron 10:587–594

Parekh SR, Cheryan M (1991) Production of acetate by mutant strains of *Clostridium thermoaceticum*. Appl Microbiol Biotechnol 36:384–387

Parrish FW (1977) Fungal transformation of 2,4-dinitrotoluene and 2,4,6-trinitrotoluene. Appl Environ Microbiol 34:232–233

Parshina SN, Sipma J, Nakashimada Y, Henstra AM, Smidt H, Lysenko AM, Lens PN, Lettinga G, Stams AJ (2005) *Desulfotomaculum carboxydivorans* sp. nov., a novel sulfate-reducing bacterium capable of growth at 100 % CO. Int J Syst Evol Microbiol 55:2159–2165

Parshina SN, Kijlstra S, Henstra AM, Sipma J, Plugge CM, Stams AJ (2005) Carbon monoxide conversion by thermophilic sulfate-reducing bacteria in pure culture and in co-culture with *Carboxydothermus hydrogenoformans*. Appl Microbiol Biotechnol 68:390–396

Paul D, Austin FW, Arick T, Bridges SM, Burgess SC, Dandass YS, Lawrence ML (2010) Genome sequence of the solvent-producing bacterium *Clostridium carboxidivorans* strain P7T. J Bacterial 192:5554–5555

Persson B, Lan HL, Gorton L, Okamoto V, Hale PD, Boguslavsky LI, Skotheim T (1993) Amperometric biosensor based on electrocatalytic regeneration of NAD+ at redox polymer modified electrodes. Biosens Bioelectron 8:81–88

Peterson JI, Vurek GG (1984) Fiber optic sensors for biomedical applications. Science 224:123–127

Phillips JR, Clausen EC, Gaddy JL (1994) Synthesis gas as substrate for the biological production of fuels and chemicals. Appl Biochem Biotechnol 45–46:145–157

Phillips J, Klasson K, Clausen E, Gaddy J (1993) Biological production of ethanol from coal synthesis gas. Appl Biochem Biotechnol 39–40:559–571

Pierce E, Xie G, Barabote RD, Saunders E, Han CS, Detter JC, Richardson P, Brettin TS, Das A, Ljungdahl LG, Ragsdale SW (2008) The complete genome sequence of *Moorella thermoacetica* (f. *Clostridium thermoaceticum*). Environ Microbiol 10:2550–2573

Plecha S, Hall D, Tiquia-Arashiro SM (2013) Screening for novel bacteria from the bioenergy feedstock switchgrass (*Panicum virgatum* L.). Environ Technol 34:1896–1904

Poehlein A, Schmidt S, Kaster AK, Goenrich M, Vollmers J, Thürmer A, Bertsch J, Schuchmann K, Voigt B, Hecker M, Daniel R, Thauer RK, Gottschalk G, Müller V (2012) An ancient pathway combining carbon dioxide fixation withthe generation and utilization of a sodium ion gradient for ATP synthesis. PLoS One 7:e33439

Preuss A, Fimpel J, Diekert G (1993) Anaerobic transformation of 2,4,6-trinitrotoluene (TNT). Arch Microbiol 159:345–353

Qureshi N, Ezeji TC (2008) Butanol 'a superior biofuel', production from agricultural residues (renewable biomass): Recent progress in technology. Biofuels Bioprod Biorefining 2:319–330

Quinn R, Mebrahtu T, Dahl TA, Lucrezi FA, Toseland BA (2004) The role of arsine in the deactivation of methanol synthesis catalysts. Appl Catal A Gen 264:103–109

Rabou LP, Zwart LM, Vreugdenhil RWR, Bos L (2009) Tar in biomass producer gas, the Energy Research Centre of the Netherlands (ECN) experience: An enduring challenge. Energy Fuels 23:6189–6198

Rabus R, Hansen TA, Widdel F (2006) Dissimilatory sulfate- and sulfur-reducing prokaryotes. In: Dworkin M, Falkow S, Rosenberg H, Schleifer KH, Stackebrandt E (eds) The prokaryotes—a handbook on the biology of bacteria. Springer, New York, pp 659–768

Ragsdale SW (1997) The eastern and western branches of the wood/ljungdahl pathway: how the east and west were won. BioFactors 6:3–11

Ragsdale SW (2004) Life with carbon monoxide. Crit Rev Biochem Mol Biol 39:165–195

Ragsdale SW, Kumar M (1996) Nickel-Containing Carbon Monoxide Dehydrogenase/Acetyl-CoA Synthase Chem Rev 96:2515–2540

Ragsdale SW, Pierce E (2008) Acetogenesis and wood-ljungdahl pathway of $CO_2$ fixation. Biochimica et Biophysica Acta 1784:1873–1898

Ragsdale SW, Ljungdahl LG, DerVartanian DV (1983) Isolation of carbon monoxide dehydrogenase from *Acetobacterium woodii* and comparison of its properties with those of the *Clostridium thermoaceticum* enzyme. J Bacteriol 155:1224–1237

Rajagopalan S, Datar RP, Lewis RS (2002) Formation of ethanol from carbon monoxide via a new microbial catalyst. Biomass Bioenergy 23:487–493

Ramey DE (2007) Butanol: The other alternative fuel. In: A Eaglesham, RWF Hardy (eds) Agricultural biofuels: Technology, sustainability and profitability, NABC Report 19, p 264

Ratliff M, Zhu W, Deshmukh R, Wilks A, Stojiljkovic I (2001) Homologues of neisserial heme oxygenase in gram-negative bacteria: degradation of heme by the product of the *pig*A gene of p*seudomonas aeroginosa*. J Bact 183:6394–6403

Razvi A, Scholtz JM (2006) Lessons in stability from thermophilic proteins. Protein Sci 15:1569–1578

Reed WM, Bogdan ME (1985) Application of cell recycling to continuous fermentative acetic acid production. Biotech Bioeng Symp 15:641–647

Rinaldi A, Mecheri B, Garavaglia V, Licoccia S, Di Nardo P, Traversa E (2008) Engineering materials and biology to boost performance of microbial fuel cells: a critical review. Energy Environment Sci 1:417–429

Rismani-Yazdi H, Christy AD, Dehority BA, Morrison M, Yu Z, Tuovinen OH (2007) Electricity generation from cellulose by rumen microorganisms in microbial fuel cells. Biotechnol Bioeng 97:1398–1407

Rismani-Yazdi H, Carver SM, Christy AD, Tuovinen OH (2008) Cathodic limitations in microbial fuel cells: An overview. J Power Source 180:683–694

Roberts SB, Gowen CM, Brooks JP, Fong SS (2010) Genome-scale metabolic analysis of *Clostridium thermocellum* for bioethanol production. BMC Syst Biol 4:31–35

Rother M, Metcalf WW (2004) Anaerobic growth of *Methanosarcina acetivorans* C2A on carbon monoxide: an unusual way of life for a methanogenic archaeon. Proc Natl Acad Sci USA 101:16929–16934

Rother M, Oelgeschlager E, Metcalf WW (2007) Genetic and proteomic analyses of CO utilization by *Methanosarcina acetivorans*. Arch Microbiol 188:463–472

Sakai S, Nakashimada Y, Yoshimoto H, Watanabe S, Okada H, Nishio N (2004) Ethanol production from $H_2$ and $CO_2$ by a newly isolated thermophilic bacterium, *Moorella* sp. HUC22-1. Biotechnol Lett 26:1607–1612

Sakai S, Nakashimada Y, Inokuma K, Kita M, Okada H, Nishio N (2005) Acetate and ethanol production from $H_2$ and $CO_2$ by *Moorella* sp. using a repeated batch culture. J Biosci Bioeng 99:252–258

Sakai S, Imachi H, Sekiguchi Y, Tseng I, Ohashi A, Harada H, Kamagata Y (2009) Cultivation of methanogens under low-hydrogen conditions by using the coculture method. Appl Environmental Microbiol 75:4892–4896

Sapra R, Bagramyan K, Adams MWW (2003) A simple energy-conserving system: proton reduction coupled to proton translocation. Proc Nat Acad Sci USA 100:7545–7550

Sato M, Matsuura K, Fujii T (2001) Ethanol separation from ethanol-watersolution by ultrasonic atomization and its proposed mechanism based on parametric decay instability of capillary wave. J Chem Physics 114:2382–2386

Savage MD, Drake HL (1986) Adaptation of the acetogen *Clostridium thermoautotrophicum* to minimal medium. J Bacteriol 165:315–318

Savage MD, Wu ZG, Daniel SL, Lundie LL, Drake HL (1987) Carbon monoxide-dependent chemolithotrophic growth of *Clostridium thermoautotrophicum*. Appl Environ Microbiol 53:1902–1906

Saxena J, Tanner RS (2011) Effect of trace metals on ethanol production from synthesis gas by the ethanologenic acetogen, *Clostridium ragsdalei*. J Ind Microbial Biotechnol 38:513–521

Saxena J, Tanner RS (2012) Optimization of a corn steep medium for production of ethanol from synthesis gas fermentation by *Clostridium ragsdalei*. World J Microb Biotechnol 28:1553–1561

Schackmann A, Müller R (1991) Reduction of nitroaromatic compounds by different *Pseudomonas* species under aerobic conditions. Appl Mlcrobiol Biotechnol 34:809–813

Schlegel HG (1966) Physiology and biochemistry of knallgasbacteria. Adv Comp Physiol Biochem 2:185–236

Schiel-Bengelsdorf B, Dürre P (2012) Pathway engineering and synthetic biology using acetogens. FEBS Lett 586:2191–2198

Schink B (1984) *Clostridium magnum* sp nov, a non-autotrophic homoacetogenic bacterium. Arch Microbiol 137:250–255

Schmidt S, Biegel E, Müller V (2009) The ins and outs of $Na^+$ bioenergetics in *Acetobacterium woodii*. Biochim Biophys Acta 1787:691–696

Scott K, Hughes R (1996) Industrial membrane separation technology. Springer Science & Business Media, New York, p 305

Seitz WR (1987) Optical sensors based on immobilized reagents. In: Turner APF, Karube I, Wilson GS (eds) Biosensors: fundamentals and applications. Oxford University Press, London, pp 599–616

Senillou A, Jaffrezic-Renault A, Martelet C, Cosnier S (1999) A miniaturized urea sensor based on the integration of both ammonium based urea enzyme field effect transistor and a reference field effect transistor in a single chip. Talanta 50:219–226

Seravalli J, Zhao S, Ragsdale SW (1999) Mechanism of transfer of the methyl group from (6S)-methyltetrahydrofolate to the corrinoid/iron-sulfur protein catalyzed by the methyltransferase from *Clostridium thermoaceticum*: a key step in the wood-ljungdahl pathway of acetyl-CoA synthesis. Biochemistry 38:5728–5735

Shanmugasundaram T, Wood HG (1992) Interaction of ferredoxin with carbon monoxide dehydrogenase from *Clostridium thermoaceticum*. J Biol Chem 267:897–900

Sharak Genthner BR, Bryant MP (1982) Growth of *Eubacterium limosum* with carbon monoxide as the energy source. Appl Environ Microbiol 43:70–74

Shaw AJ, Podkaminer KK, Desai SG, Bardsley JS, Rogers SR, Thorne PG, Hogsett DA, Lynd LR (2008) Metabolic engineering of a thermophilic bacterium to produce ethanol at high yield. Proc Natl Acad Sci USA 105:13769–13774

Shima S, Lyon EJ, Sordel-Klippert MS, Kauss M, Kahnt J, Thauer RK, Steinbach K, Xie XL, Verdier L, Griesinger C (2004) The cofactor of the iron-sulfur cluster free hydrogenase Hmd: structure of the light-inactivation product. Angewandte Chemie-International Edition 43:2547–2551

Shen GJ, Shieh JS, Grethlein AJ, Jain MK, Zeikus JG (1999) Biochemical basis for carbon monoxide tolerance and butanol production by *Butyribacterium methylotrophicum*. Appl Microbiol Biotechnol 51:827–832

Shewa WE, Chaganti SR, Lalman JA (2014) Electricity generation and biofilm formation in microbial fuel cells using plate anodes constructed from various grades of graphite. J Green Eng 4:13–32

Siebers B, Schönheit P (2005) Unusual pathways and enzymes of central carbohydrate metabolism in archaea. Curr Opin Microbiol 8:695–705

Siedlecki M, de Jong W, Verkooijen AHM (2011) Fluidized bed gasification as a mature and reliable technology for the production of bio-syngas and applied in the production of liquid transportation fuels-A review. Energies 4:389–434

Sim JH, Kamaruddin AH (2008) Optimization of acetic acid production from synthesis gas by chemolithotrophic bacterium—Clostridium aceticum using statistical approach. Bioresour Technol 99:2724–2735

Sim JH, Kamaruddin AH, Long WS, Najafpour G (2007) Clostridium aceticum—A potential organism in catalyzing carbon monoxide to acetic acid: Application of response surface methodology. Enzyme Microb Technol 40:1234–1243

Singer SW, Hirst MB, Ludden PW (2006) CO-dependent $H_2$ evolution by *Rhodospirillum rubrum*: role of CODH:CooF complex. Biochim Biophys Acta 1757:1582–1591

Sipma J, Henstra AM, Parshina SN, Lens PNL, Lettinga G, Stams AJM (2006) Microbial CO conversions with applications in synthesis gas purification and bio-desulfurization. Crit Rev Biotechnol 26:41–65

Sittig M (1985) Handbook of toxic and hazardous chemicals and carcinogens. Noyes Publications, Park Ridge, p 950

Sizova MV, Izquierdo JA, Panikov NS, Lynd LR (2011) Cellulose- and xylan-degrading thermophilic anaerobic bacteria from biocompost. Appl Environ Microbiol 77:2282–2291

Slepova TV, Sokolova TG, Lysenko AM, Tourova TP, Kolganova TV, Kamzolkina OV, Karpov GA, Bonch-Osmolovskaya EA (2006) *Carboxydocella sporoproducens* sp. nov., a novel anaerobic CO-utilizing/$H_2$-producing thermophilic bacterium from a kamchatka hot spring. Int J Syst Evol Microbiol 56:797–800

Slepova TV, Sokolova TG, Kolganova TV, Tourova TP, Bonch-Osmolovskaya EA (2009) *Carboxydothermus siderophilus* sp. nov., a thermophilic, hydrogenogenic, carboxydotrophic, dissimilatory Fe(III)-reducing bacterium from a kamchatka hot spring. Int J Syst Evol Microbiol 59:213–217

Slobodkin A, Reysenbach AL, Strutz N, Dreier M, Wiegel J (1997) *Thermoterrabacterium ferrireducens* gen. nov., sp. nov., a thermophilic anaerobic dissimilatory Fe(III)-reducing bacterium from a continental hot spring. Int J Syst Bacteriol 47:541–547

Slobodkin AI, Sokolova TG, Lysenko AM, Wiegel J (2006) Reclassification of *Thermoterrabacterium ferrireducens* as *Carboxydothermus ferrireducens* comb. nov., and emended description of the genus *Carboxydothermus*. Int J Syst Evol Micrpbiol 56:2349–2351

Slobodkina GB, Panteleeva AN, Sokolova TG, Bonch-Osmolovskaya EA, Slobodkin AI (2012) *Carboxydocella manganica* sp. nov., a thermophilic, dissimilatory Mn(IV)- and Fe(III)-reducing bacterium from a kamchatka hot spring. Int J Syst Evol Microbiol 62:890–894

Smith KD, Klasson KT, Ackerson MD, Clausen EC, Gaddy JL (1991) COS degradation by selected CO-utilizing bacteria. Appl Biochem Biotechnol 28–29:787–796

Soboh B, Linder D, Hedderich R (2002) Purification and catalytic properties of a CO-oxidizing: $H_2$-evolving enzyme complex from *Carboxydothermus hydrogenoformans*. Eur J Biochem 269:5712–5721

Soboh B, Linder D, Hedderich R (2004) A multisubunit membrane-bound [NiFe] hydrogenase and an NADH-dependent Fe-only hydrogenase in the fermenting bacterium *Thermoanaerobacter tengcongensis*. Microbiology 150:2451–2463

Sokolova TG, Lebedinsky A (2013) CO-oxidizing anaerobic thermophilic prokaryotes. T Satyanarayana, J Littlechild, Y Kawarabayasi (eds) Thermophilic Microbes in the Environment and Industrial Biotechnology. Springer Publisher, New York, p. 203–231

Sokolova TG, Gonzalez JM, Kostrikina NA, Chernyh NA, Tourova TP, Bonch-Osmolovskaya EA, Robb FT (2001) *Carboxydobrachium pacificum* gen. nov., sp.nov., a new anaerobic thermophilic carboxydotrophic bacterium from Okinawa trough. Int J Syst Bacteriol 51:141–149

Sokolova TG, Kostrikina NA, Chernyh NA, Tourova TP, Kolganova TV, Bonch-Osmolovskaya EA (2002) *Carboxydocella thermautotrophica* gen. nov., sp. nov., a novel anaerobic CO-utilizing thermophile from a kamchatkan hot spring. Int J Syst Evol Microbiol 52:1961–1967

Sokolova TG, Jeanthon C, Kostrikina NA, Chernyh NA, Lebedinsky AV, Stackebrandt E, Bonch-Osmolovskaya EA (2004) The first evidence of anaerobic CO oxidation coupled with $H_2$ production by a hyperthermophilic archaeon isolated from a deep-sea hydrothermal vent. Extremophiles 8:317–323

Sokolova TG, Gonzalez JM, Kostrikina NA, Chernyh NA, Slepova TV, Bonch-Osmolovskaya EA, Robb FT (2004) *Thermosinus carboxydivorans* gen. nov., sp. nov., a new anaerobic thermophilic, carbon-monoxide-oxidizing, hydrogenogenic bacterium from a hot pool of Yellowstone National Park. Int J Syst Evol Microbiol 54:2353–2359

Sokolova TG, Kostrikina NA, Chernyh NA, Kolganova TV, Tourova TP, Bonch-Osmolovskaya EA (2005) *Thermincola carboxydiphila* gen. nov., sp. nov., a new anaerobic carboxydotrophic hydrogenogenic bacterium from a hot spring of Lake Baikal area. Int J Syst Evol Microbiol 55:2069–2073

Sokolova T, Hanel J, Oneynwoke RU, Reysenbach A-L, Banta A, Geyer R, Gonzalez J, Whitman WB, Wiegel J (2007) Novel chemolithotrophic, thermophilic, anaerobic bacteria *Thermolithobacter ferrireducens* gen. nov., sp. nov. and *Thermolithobacter carboxydivorans* sp. nov. Extremophiles 11:145–157

Sokolova TG, Henstra AM, Sipma J, Parshina SN, Stams AJ, Lebedinsky AV (2009) Diversity and ecophysiological features of thermophilic carboxydotrophic anaerobes. FEMS Microbiol Ecol 68:131–141

Song C (2002) Fuel processing for low-temperature and high-temperature fuel cells: Challenges, and opportunities for sustainable development in the 21st century. Catalysis Today 77:17–49

Soucaille P, Figge R, Croux C (2008) Process for Chromosomal Integration and DNA Sequence Replacement in *Clostridia*. Patent WO/2008/040387

Sowers KR, Baron SF, Ferry JG (1984) *Methanosarcina acetivorans* sp. nov., an acetotrophic methane-producing bacterium isolated from marine sediments. Appl Environ Microbiol 47:971–978

Spain AM, Krumholz LR (2011) Nitrate-reducing bacteria at the nitrate and radionuclide contaminated Oak Ridge integrated field research challenge site: A review. Geomicrobiol J 28:418–429

Spath PL, Dayton DC (2003).Preliminary screening: Technical and economic assessment of synthesis gas to fuels and chemicals with emphasis on the potential for biomass-derived syngas. Technical report NREL/TP-510-34929, National Renewable Energy Lab, Golden, Colorado, p 160

Steele BCH, Heinzel A (2001) Materials for fuel-cell technologies. Nature 414:345–352

Stephanopoulos, G (2011) Bioprocess and Microbe Engineering for Total Carbon Utilization in Biofuel Production. U.S. Patent 0177564

Stetter KO (1988) *Archaeoglobus fulgidus* gen. nov., sp. nov.: a new taxon of extremely thermophilic archaebacteria. Syst Appl Microbiol 10:172–173

Stetter KO (1996) Hyperthermophilic prokaryotes. FEMS Microbiol Rev 18:149–158

Suflita JM, Horowitz A, Shelton DR, Tiedje JM (1982) Dehalogenation: a novel pathway for the anaerobic biodegradation of haloaromatic compounds. Science 218:1115–1117

Suflita JM, Robinson JA, Tiedje JM (1983) Kinetics of microbial dehalogenation of haloaromatic substrates in methanogenic environments. Appl Environ Microbiol 45:1466–1473

Suflita JM, Stout J, Tiedje JM (1984) Dechlorination of 2,4,5-trichlorophenoxyacetic acid (2,4,5-T) by anaerobic microorganisms. J Agric Food Chem 32:218–221

Sukeerthi S, Contractor AQ (1994) Applications of conducting polymers as sensors. Ind J Chem 33A:565–571

Suryawanshi PC, Chaudhari AB, Kothari RM (2010) Thermophilic anaerobic digestion: the best option for waste treatment. Crit Rev Biotechnol 30:31–40

Suzuki S, Karube I (1982) Microbial sensors for gas analysis. In: EK Pye, LB Wingard (eds) Enzyme engineering. vol. 6, Plenum Press, New York, p 387–393

Svetlichnyi VA, Sokolova TG, Kostrikina NA, Lysenko AM (1994) *Carboxydothermus restrictus* sp. nov. - a new thermophilic anaerobic carboxydotrophic bacterium. Mikrobiologiya 63:523–528

Svetlitchnyi V, Peschel C, Acker G, Meyer O (2001) Two membraneassociatedNiFeS-carbon monoxide dehydrogenases from the anaerobic carbon-monoxide-utilizing eubacterium *Carboxydothermus hydrogenoformans*. J Bacteriol 183:5134–5144

Svetlichny VA, Sokolova TA, Gerhardt M, Kostrikina NA, Zavarzin GA (1991) Anaerobic extremely thermophilic carboxydotrophic bacteria in hydrotherms of Kuril islands. Microb Ecol 21:1–10

Svetlichny VA, Sokolova TG, Gerhardt M, Ringpfeil M, Kostrikina NA, Zavarzin GA (1991) *Carboxydothermus hydrogenoformans*, gen. nov., spec. nov., a carbon monoxide utilizing thermophilic anaerobic bacterium from hydrothermal environments of Kunashir island. Syst Appl Microbiol 14:254–260

Svetlichny VA, Sokolova TG, Kostrikina NA, Lysenko AM (1994) A new thermophilic anaerobic carboxydotrophic bacterium *Carboxydothermus restrictus* sp. nov. Microbiology 63:294–297

Symonds RB, Rose WI, Bluth G, Gerlach TM (1994) Volcanic gas studies: methods, results, and applications. In: MR Carroll, JR Holloway (eds) Volatiles in magmas. Mineralogical Society of America Reviews in Mineralogy, p 1–66

Takeguchi T, Yanagisawa KI, Inui T, Inoue M (2000) Effect of the property of solid acid upon syngas-to-dimethyl ether conversion on the hybrid catalysts composed of Cu-Zn-Ga and solid acids. Appl Catalysis A Gen. 192:201–209

Tanner RS, Miller LM, Yang D (1993) *Clostridium ljungdahlii* sp. nov., an acetogenic species in clostridial rRNA homology group 1. Int J Syst Bacteriol 43:232–236

Taylor MP, Eley KL, Martin S, Tuffin MI, Burton SG, Cowan DA (2009) Thermophilic ethanologenesis: future prospects for second-generation bioethanol production. Trends Biotechnol 27:398–405

Techtmann SM, Colman AS, Robb FT (2009) 'That which does not kill us only makes us stronger': the role of carbon monoxide in thermophilic microbial consortia. Environ Microbiol 11:1027–1037

Terlesky KC, Nelson MJK, Ferry JG (1986) Isolation of an enzyme complex with carbon monoxide dehydrogenase activity containing corrinoid and nickel from acetategrown *Methanosarcina thermophila*. J Bacteriol 168:1053–1058

Thauer RK (2007) Microbiology - A fifth pathway of carbon fixation. Science 318:1732–1733

Thauer RK (1998) Biochemistry of methanogenesis: a tribute to marjory stephenson. Microbiology 144:2377–2406

Tierney LM (2004) Current medical diagnosis and treatment. McGraw-Hill, USA 1818 p

Tiquia SM (2010) Salt-adapted bacteria isolated from the Rouge River and potential for degradation of contaminants and biotechnological applications. Environ Technol 31:967–978

Tiquia SM, Mormile MR (2010) Extremophiles–A source of innovation for industrial and environmental applications. Editorial overview. Environ Technol 31:823

Tiquia-Arashiro SM, Mormile MR (2013) Sustainable technologies: bioenergy and biofuel from biowaste and biomass. Editorial Overview. Environ Technol 34:1637–1638

Tirado-Acevedo O, Chinn MS, Grunden AM (2010) Production of biofuels from synthesis gas using microbial catalysts. Adv Appl Microbiol. 70:57–92

Tor JM, Kashefi K, Lovley DR (2001) Acetate oxidation coupled to Fe (III) reduction in hyperthermophilic microorganisms. Appl Environ Microbiol 67:1363–1365

Tracy B, Papoutsakis E (2010) Methods and Compositions for Genetically Engineering *Clostridia* Species. U.S. Patent 20120301964 A1

Tracy BP, Jones SW, Papoutsakis ET (2011) Inactivation of σE and σG in *Clostridium acetobutylicum* illuminates their roles in clostridial-cell-form biogenesis, granulose synthesis, solventogenesis, and spore morphogenesis. J Bacterial 193:1414–1426

Tracy BP, Jones SW, Fast AG, Indurthi DC, Papoutsakis ET (2012) *Clostridia*: the importanceof their exceptional substrate and metabolite diversity for biofuel and biorefinery applications. Curr Opin Biotechnol 23:364–381

Tripathi S, Olson DG, Argyros DA, Miller BB, Barrett TF, Murphy DM, McCool JD, Warner AK, Rajgarhia VB, Lynd LR, Hogsett DA, Caiazza NC (2010) Development of pyrF-based genetic system for targeted gene deletion in *Clostridium thermocellum* and creation of a pta mutant. Appl. Environ. Microbiol 76:6591–6599

Tummala SB, Welker NE, Eleftherios T (1999) Development and characterization of a gene expression reporter system for *Clostridium acetobutylicum* ATCC 824. Appl Environ Microbiol 65:3793–3799

Turner APF, Aston WJ, Higgins IJ, Davis G, Hill HAO (1982) Applied aspects of bioelectrochemistry: fuel cells, sensors, and bioorganic synthesis. Biotechnol Bioeng Symp 12:401–412

Turner APF, Ramsey G, Higgins IJ (1983) Applications of electron transfer between biological systems and electrodes. Biochem Soc Trans 11:445–448

Turner APF, Aston WJ, Davis G, Higgins IJ, Hill HAO, Colby J (1985) Enzyme-based carbon monoxide sensors. In: RK Poole, CS Dow (eds) Microbial gas metabolism: Mechanistic, metabolic and biotechnological aspects. Special publications of the Society for General Microbiology, Academic Press, New York, pp 161–170

Turner A, Wilson G, Kaube I (1987) Biosensors: fundamentals and applications. Oxford University Press, Oxford 770 p

Turner P, Mamo G, Karlsson EN (2007) Potential and utilization of thermophiles and thermostable enzymes in biorefining. Microb Cell Fact 15:6–9

Tyagi M, da Fonseca MM, de Carvalho CC (2011) Bioaugmentation and biostimulation strategies to improve the effectiveness of bioremediation processes. Biodegradation 22:231–241

Uffen RL (1976) Anaerobic growth of a *Rhodopseudomonas* species in the dark with carbon monoxide as sole carbon and energy substrate. Proc Natl Acad Sci USA 73:3298–3302

Uffen RL (1981) Metabolism of carbon monoxide. Enzyme Microb Technol 3:197–206

Ugwu CU, Ogbonna JC (2002) Improvement of mass transfer characteristics and productivities of inclined tubular photobioreactors by installation of internal static mixers. Appl Microbiol Biotechnol 58:600–607

Ukpong MN, Atiyeh HK, De Lorme MJM, Liu K, Zhu X, Tanner RS, Wilkins MR, Stevenson BS (2012) Physiological response of *Clostridium carboxidivorans* during conversion of synthesis gas to solvents in a gas-fed bioreactor. Biotechnol Bioeng 109:2720–2728

van der Drift A, van Doorn J, Vermeulen JW (2001) Ten residual biomass fuels for circulating fluidized-bed gasification. Biomass Bioener 20:45–56

Vega J, Prieto S, Elmore B, Clausen E, Gaddy J (1989) The Biological production of ethanol from synthesis gas. Appl Biochem Biotechnol 20–21:781–797

Vega JL, Clausen EC, Gaddy JL (1990) Design of bioreactors for coal synthesis gas fermentations. Resour Conserv Recyc 3:149–160

Vieille C, Zeikus GJ (2001) Hyperthermophilic enzymes: sources, uses, and molecular mechanisms for thermostability. Microbiol Mol Biol Rev 65:1–43

Vignais PM, Billoud B, Meyer J (2001) Classification and phylogeny of hydrogenases. FEMS Microbiol Rev 25:455–501

Vo-Dinh T, Cullum B (2000) Biosensors and biochips: Advances in biological and medical diagnostics. Fresenius' J Anal Chem 366:540–551

Vogel TM, Criddle CS, McCarty PL (1987) Transformations of halogenated aliphatic compounds. Environ Sci Technol 21:722–736

Vorapongsathorn T, Wongsuchoto P, Pavasant P (2001) Performance of airlift contactors with baffles. Chem Eng J 84:551–556

Wagner FS Jr (2002) Acetic Acid. Kirk-Othmer Encycl Chem Technol 1:115–136

Wagner AO, Malin C, Lins P, Illmer P (2011) Effects of various fatty acid amendments on a microbial digester community in batch culture. Waste Manage 31:431–437

Wasserfallen A, Nolling J, Pfister P, Reeve J, Conway de Macario E (2000) Phylogenetic analysis of 18 thermophilic *Methanobacterium* isolates supports the proposals to create a new genus, *Methanothermobacter* gen. nov., and to reclassify several isolates in three species, *Methanothermobacter thermautotrophicus* comb. nov., *Methanothermobacter wolfeii* comb. nov., and *Methanothermobacter marburgensis* sp. nov. Int J Syst Evol Microbiol 50:43–53

Watanabe T, Asakawa S, Nakamura A, Nagaoka K, Kimura M (2004) DGGE method for analyzing 16S rDNA of methanogenic archaeal community in paddy field soil. FEMS Microbiol Lett 232:153–163

Watson J (1984) The tin oxide gas sensor and its applications. Sensors Actuators 5:29–42

Weaver JC, Cooney CL, Fulton SP, Schuler D, Tannenbaum SR (1976) Experiments and calculation concerning a thermal enzyme probe. Biochim Biophys Acta 452:258–291

Weber KA, Achenbach LA, Coates JD (2006) Microorganisms pumping iron: anaerobic microbial iron oxidation and reduction. Nat Rev Microbiol 4:752–764

Wei J, Liang P, Huang X (2011) Recent progress in electrodes for microbial fuel cells. Biores Technol 102:9335–9344

Wei L, Pordesimo LO, Haryanto A, Wooten J (2011) Co-gasification of hardwood chips and crude glycerol in a pilot scale downdraft gasifier. Biores Technol 102:6266–6272

Wetchakun K, Samerjai T, Tamaekong N, Liewhiran C, Siriwong C, Kruefu V, Wisitsoraat A, Tuantranont A, Phanichphant S (2011) Semiconducting metal oxides as sensors for environmentally hazardous gases. Sensors Actuators. 160:580–591

White H, Lebertz H, Thanos I, Simon H (1987) *Clostridium thermoaceticum* forms methanol from carbon monoxide in the presence of viologen dyes. FEMS Microb Lett 43:173–176

Wiegel J (1980) Formation of ethanol by bacteria. A pledge for the use of extreme thermophilic anaerobic bacteria in industrial ethanol fermentation processes. Experientia 36:1434–1446

Wiegel J, Ljungdahl LG (1986) The importance of thermophilic bacteria in biotechnology. CRS Rev Biotechnol 3:38–108

Wiegel J, Braun M, Gottschalk G (1981) *Clostridium thermoautotrophicum* species novum, a thermophile producing acetate from molecular hydrogen and carbon dioxide. Curr Microbiol 5:255–260

Wiegel J, Carreira LH, Garrison R, Rabek NE, Ljungdahl LG (1991) Calcium magnesium acetate (CMA) manufacture from glucose by fermentation with thermophilic homoacetogenic bacteria. In: Wise DL, Lavendis YA, Metghalchi M (eds) Calcium magnesium acetate. Elsevier Science Publisher, Amsterdam, pp 359–418

Wieringa KT (1936) Over het verdwijnen van waterstof en koolzuur onder anaerobe voorwaarden. Antonie van Leeuwenhoek 3:263–273

Wieringa KT (1939) The formation of acetic acid from carbon dioxide and hydrogen by anaerobic spore-forming bacteria. Antonie van Leeuwenhoek 6:251–262

Wilhelm DJ, Simbeck DR, Karp AD, Dickenson RL (2001) Syngas production for gas-to-liquids applications: Technologies, issues and outlook. Fuel Process Technol 71:139–148

Willems A, Gillis M. DeLey J (1991) Transfer of *Rhodocyclus gelatinosus* to *Rubrivivax gelatinosus* gen. nov., comb. nov., and phylogenetic relationships with *Leptothrix, Sphaerotilus natans, Pseudomonas saccharophila*, and *Alcaligenes latus*. Int J Syst Bact 41: 65–73

Williesa S, Isupova M, Littlechild J (2010) Thermophilic enzymes and their applications in biocatalysis: a robust aldo-keto reductase. Environ Technol 31:1159–1167

Winter J, Lerp C, Zabel HP, Wildenauer FX, König H, Schindler F (1984) *Methanobacterium wolfei*, sp. nov., a new tungsten-requiring, thermophilic, autotrophic methanogen. Syst Appl Microbiol 5:457–466

Worden RM, Grethlein AJ, Jain MK, Datta R (1991) Production of butanol and ethanol from synthesis gas via fermentation. Fuel 70:615–619

Worden RM, Grethlein AJ, Zeikus JG, Datta R (1989) Butyrate production from carbon monoxide by *Butyribacterium methylotrophicum*. Appl Biochem Biotechnol 20–21:687–698

Wood HG (1991) Life with CO or $CO_2$ and $H_2$ as a source of carbon and energy. FASEB J 5:156–163

Wood HG, Ragsdale SW, Pezacka P (1986) The acetyl-CoA pathway of autotrophic growth. FEMS Microbiol Lett 39:345–362

Woods DR (1995) The genetic engineering of microbial solvent production. Trends Biotechnol 13:259–264

Wrighton KC, Coates JD (2009) Microbial Fuel Cells: Plug-in and Power-on Microbiology. Microbe 4:281–287

Wrighton KC, Agbo P, Warnecke F, Weber KA, Brodie EL, DeSantis TZ, Hugenholtz P, Andersen GL, Coates JD (2008) A novel ecological role of the *Firmicutes* identified in thermophilic microbial fuel cells. ISME J 2:1146–1156

Wu M, Ren Q, Durkin AS, Daugherty SC, Brinkac LM, Dodson RJ, Madupu R, Sullivan SA, Kolonay JF, Haft DH, Nelson WC, Tallon LJ, Jones KM, Ulrich LE, Gonzalez JM, Zhulin IB, Robb FT, Eisen JA (2005) Life in hot carbon monoxide: the complete genome sequence of *Carboxydothermus hydrogenoformans* Z-2901. PLOS Genet 1:563–574

Wu WM, Gu B, Fields MW, Gentile M, Ku YK, Yan H, Tiquias S, Yan T, Nyman J, Zhou J, Jardine PM, Criddle CS (2005) Uranium (VI) reduction by denitrifying biomass. Bioremediation J 9:49–61

Xu CC, Donald J, Byambajav E, Ohtsuka Y (2010) Recent advances in catalysts for hot-gas removal of tar and NH$_3$ from biomass gasification. Fuel 89:1784–1795

Yamada T, Suzuki T (1983) Occurrence of reductive dechlorination products in the paddy field soil treated with CNP (chloronitrofen). J Pestic Sci 8:437–443

Yoneda Y, Yoshida T, Kawaichi S, Daifuku T, Takabe K, Sako Y (2012) *Carboxydothermus pertinax* sp. nov., a thermophilic, hydrogenogenic, Fe(III)-reducing, sulfur-reducing carboxydotrophic bacterium from an acidic hot spring. Int J Syst Evol Microbiol 62:1692–1697

Yoneda Y, Yoshida T, Yasuda H, Imada C, Sako Y (2013) A thermophilic, hydrogenogenic and carboxydotrophic bacterium, *Calderihabitans maritimus* gen. nov., sp. nov., from a marine sediment core of an undersea caldera. Int J Syst Evol Microbiol 63:3602–3628

Younesi H, Najafpour G, Ku Ismail KS, Mohamed AR, Kamaruddin AH (2008) Biohydrogen production in a continuous stirred tank bioreactor from synthesis gas by anaerobic photosynthetic bacterium: *Rhodopirillumrubrum*. Biores Technol 99:2612–2619

Younesi H, Najafpour G, Mohameda AR (2005) Ethanol and acetate production from synthesis gas via fermentation processes using anaerobic bacterium, Clostridium ljungdahlii. Biochem Engineer J 27:110–119

Zavarzin GA, Nozhevnikova AN (1977) Aerobic carboxydobacteria. Microb Ecol 3:305–326

Zavarzina DG, Sokolova TG, Tourova TP, Chernyh NA, Kostrikina NA, Bonch-Osmolovskaya EA (2007) *Thermincola ferriacetica* sp. nov., a new anaerobic, thermophilic, facultatively chemolithoautotrophic bacterium capable of dissimilatory Fe (III) reduction. Extremophiles 11:1–7

Zeikus JG (1983) Metabolism of one-carbon compounds by chemotrophic anaerobes. Adv Microbial Physiology 24:215–299

Zeikus JG, Lynd LH, Thompson TE, Krzycki JA, Weimer PJ, Hegge PW (1980) Isolation and characterization of a new, methylotrophic, acidogenic anaerobe, the Marburg strain. Curr Microbiol 3:381–386

Zellner G, Stackebrandt E, Kneifel H, Messner P, Sleytr UB, Demacario EC, Zabel HP, Stetter KO, Winter J (1989) Isolation and characterization of a thermophilic, sulfate-reducing archaebacterium, *Archaeoglobus fulgidus* strain-Z. Syst Appl Microbiol 11:151–160

Zeng S, Baillargeat D, Ho HP, Yong KT (2014) Nanomaterials enhanced surface plasmon resonance for biological and chemical sensing applications. Chem Soc Rev 43:3426–3452

Zhang P, Liu Z (2010) Experimental study of the microbial fuel cell internal resistance. J Power Sources 195:8013–8018

Zinder SH, Mah RA (1979) Isolation and characterization of a thermophilic strain of *Methanosarcina* unable to use H$_2$-CO$_2$ for methanogenesis. Appl Environ Microbiol 38:996–1008

Zinder SH, Sowers KR, Ferry JG (1985) *Methanosarcina thermophila* sp. nov. a thermophilic, acetotrophic, methane-producing bacterium. Int J Syst Bacteriol 35:522–523

Zirngibl C, Vandongen W, Schworer B, Vonbunau R, Richter M, Klein A, Thauer RK (1992) H$_2$-Forming methylenetetrahydromethanopterin dehydrogenase, a novel type of hydrogenase without iron-sulfur clusters in methanogenic archaea. Eur J Biochem 208:511–520

Zuo Y, Jones RD (1995) Formation of carbon-monoxide by photolysis of dissolved marine organic material and its significance in the carbon cycling of the oceans. Naturwissenschaften 82:472–474